PPM

Practical Problems in Mathematics

FOR RENEWABLE ENERGY TECHNICIANS

Practical Problems in Mathematics

FOR RENEWABLE ENERGY TECHNICIANS

Russell B. DeVore

CENGAGE
Learning®

Australia • Brazil • Mexico • Singapore • United Kingdom • United States

CENGAGE
Learning®

**Practical Problems in Mathematics
for Renewable Energy Technicians**
Russell B. DeVore

SVP, GM Skills & Global Product
Management: Dawn Gerrain

Product Director: Matthew Seeley

Product Team Manager: James DeVoe

Product Manager: Vanessa Myers

Senior Director, Development:
Marah Bellegarde

Senior Product Development Manager:
Larry Main

Senior Content Developer: Mary Clyne

Product Assistant: Jason Koumourdas

Vice President, Marketing Services:
Jennifer Ann Baker

Marketing Director: Michele McTighe

Senior Production Director: Wendy Troeger

Production Director: Andrew Crouth

Senior Content Project Manager:
James Zayicek

Content Project Management and Art
Direction: Lumina Datamatics, Inc.

Cover image(s): Vaclav Volrab/Shutterstock

Unless otherwise noted, all items
© Cengage Learning

For product information and technology assistance, contact us at
Cengage Learning Customer & Sales Support, 1-800-354-9706

For permission to use material from this text or product,
submit all requests online at **www.cengage.com/permissions.**
Further permissions questions can be e-mailed to
permissionrequest@cengage.com

Library of Congress Control Number: 2015947219

ISBN: 978-1-2850-7933-2

Cengage Learning
20 Channel Center Street
Boston, MA 02210
USA

Cengage Learning is a leading provider of customized learning solutions
with employees residing in nearly 40 different countries and sales in
more than 125 countries around the world. Find your local representative
at **www.cengage.com**.

Cengage Learning products are represented in Canada by
Nelson Education, Ltd.

To learn more about Cengage Learning, visit **www.cengage.com**

Purchase any of our products at your local college store or at our
preferred online store **www.cengagebrain.com**

Notice to the Reader
Publisher does not warrant or guarantee any of the products described herein or perform any inde-
pendent analysis in connection with any of the product information contained herein. Publisher does
not assume, and expressly disclaims, any obligation to obtain and include information other than
that provided to it by the manufacturer. The reader is expressly warned to consider and adopt all
safety precautions that might be indicated by the activities described herein and to avoid all potential
hazards. By following the instructions contained herein, the reader willingly assumes all risks in connection with such instructions. The publisher makes no representations or warranties of any kind,
including but not limited to, the warranties of fitness for particular purpose or merchantability, nor
are any such representations implied with respect to the material set forth herein, and the publisher
takes no responsibility with respect to such material. The publisher shall not be liable for any special,
consequential, or exemplary damages resulting, in whole or part, from the readers' use of, or reliance
upon, this material.

Printed in the United States of America
Print Number: 01 Print Year: 2015

To my support system and inspiration—Fran

CONTENTS

PREFACE

As a student I found that the easiest way to learn a subject was to relate it to something of interest to me. If I could see the relevance to a topic that I enjoyed, I could spend time on the subject without losing interest. In fact, many times, if a difficult subject was related to one that I liked, it would seem easy to me. This is the philosophy behind the Practical Problems series. I have found that my students understand the concepts better if they can relate to the application of the topic.

Throughout the development of *Practical Problems in Mathematics for Renewable Energy Technicians,* I have tried hard to make each problem relevant to the fast-developing field of renewable energy technologies. You might not encounter every application in your work experiences, but each of you will encounter many of these applications. It is my hope that this text will make the learning of math easier.

CENGAGE LEARNING'S PPM SERIES

This text is one of a series of workbooks designed to offer students practical problem-solving experience in various occupations. The workbooks take a step-by-step approach to mastering basic math skills. Each workbook includes relevant and easily understood problems in a specific vocational field. The workbooks are suitable for any student from junior high through high school and up to the two-year college level. Each text includes a glossary to help students with technical terms. *Practical Problems in Mathematics for Renewable Energy Technicians* includes an appendix with information on English and SI measurements, important formulas, and answers to odd-numbered questions. For more information about this series and a current list of titles, please visit www.cengagebrain.com.

SERIES FEATURES

The workbooks in Cengage Learning's PPM series take a step-by-step approach to mastering essential math skills. At the start of each unit, a brief introductory section provides a basic explanation of the concepts necessary to complete the problems in the unit. Examples are presented to help the learner review the mathematical principles. The problems in each unit progress from basic examples of the math concepts to more complex examples that require critical thinking. As students progress through each unit, they will become more proficient at solving a wide variety of math problems.

THIS BOOK'S APPROACH

Practical Problems in Mathematics for Renewable Energy Technicians begins with a review of basic operations with whole numbers, fractions, and decimals and progresses through ratio and proportion, measurements, and formulas to finish with sections on trigonometry, graphs, and business topics. Topical sections are divided into short units to give teachers maximum flexibility in planning and to help students achieve maximum skill mastery. Instructors may choose to use this book as a stand-alone text or as a supplemental workbook to a theory-based text.

INSTRUCTOR RESOURCES

The Instructor Companion Website provides the following support for teachers:

- Answers to all text problems
- PowerPoint presentations
- An Image Gallery including all text figures
- Cengage Learning Testing Powered by Cognero

Cengage Learning Testing Powered by Cognero is a flexible, online system that allows you to:

- author, edit, and manage test bank content from multiple Cengage Learning solutions
- create multiple test versions in an instant
- deliver tests from your LMS, your classroom, or wherever you want

ACKNOWLEDGMENTS

The publisher and author would like to thank the following reviewers for the valuable suggestions during the development of this edition:

Content Review:

Debbie French
New Philadelphia High School
New Philadelphia, Ohio

Technical Review:

Linda Willey
Clifton Park, New York

ABOUT THE AUTHOR

Russell B. DeVore was chair of the Arts & Sciences Division at Trident Technical College in Charleston, South Carolina. He then taught in the Physics Department at Bloomsburg University in Pennsylvania. Dr. DeVore was named in the 1976 edition of *Who's Who in South Carolina and Outstanding Young Men of America*. He is retired, having served as a shift technical advisor and simulator instructor at the training center of a nuclear power plant for PPL Susquehanna, LLC. He is currently an adjunct professor teaching nuclear power plant science at Luzerne County Community College and a mentor for Thomas Edison State College.

INTRODUCTION

Before beginning this text, take some time to review the three topics discussed in this introduction. This information may help you to make fewer errors when working math problems and may also make working the problems easier.

CALCULATORS

Today, the availability and low cost of calculators have made many math problems less of a chore. Modern calculators can do many calculations quickly. Thankfully, a complicated math calculation is no longer a big problem.

However, calculators can also have a "bad" side to them. Some people rely on them to do all of their math; they have gotten lazy when doing math. These people may get into trouble for a number of reasons.

First, when they are doing a math problem, these people have no idea if they have done it correctly. If an incorrect number was entered, or an incorrect button was pressed, these people would never know because they have no idea what kind of answer to expect. A couple of examples might clarify this point:

1. A person who relies totally on the calculator for math problems is multiplying 8.4×2.1. The calculator gives 4 as the answer. This person does not realize that a mistake was made by pressing the $\div$ button instead of the $\times$ button, because he or she does not know that the correct answer should be about 16.

2. Someone doing a problem may not realize that a decimal point was not pressed or that a zero did not get entered into the calculator correctly when the number was being input because, again, the person has no idea what the correct answer should be. So, multiplying 8.4×2.1 may give an answer of 176.4 rather than 17.64, and the person doing the problem may not know that an error has been made.

So it is very important when working math problems with a calculator to have a rough idea what the answer should be.

There is a second drawback to calculators—they cannot work in fractions. If people rely on calculators all of the time, they will have no idea how to solve problems involving fractions because they have not practiced. Solving math problems is largely a matter of practicing; without practice, problems are hard to solve. Some people use the calculator to convert fractions to decimals; however, some fractions cannot be converted exactly. This means that the calculator answer will not be exactly correct. Because many dimensions in the renewable energy field involve fractions, a student should be able to work with fractions.

A third drawback to calculators is that they need power, and they may run out of power at the worst possible time. A backup system should be ready, and that may be doing the math by hand.

Yet even with these drawbacks, calculators can be a big help in math. They can solve all problems that use decimals and are really helpful when there are many numbers to add, subtract, or multiply. It is important to follow some basic rules:

1. Get in the habit of thinking about what a reasonable answer would be. This makes it easier to notice when an error has been made.

2. After entering the number into the calculator, check to see that you entered the correct number before pushing the operation button.

If you are going to buy a calculator, you will probably not need many of the special keys that some calculators have on them. The extra keys often make the calculators more expensive and may be a waste of money. Having more keys does not mean that the calculator is more accurate.

Each calculator comes with an instruction booklet. Use it. Read it completely. This is the only way that you will know exactly what your calculator will or will not do. One of the biggest shortcuts when working with most calculators is that when doing multiple operations, such as adding many numbers, you do not have to press the = button after each number is entered. Pressing the + button will cause the calculator to add the number just entered to the total already in the calculator and be ready to have another number entered. You have to press the = button only once, at the very end, to get the final sum. This same idea works for subtracting and multiplying.

Remember that not all calculators work this way. Check the instruction booklet to make sure your calculator has this feature.

Calculators can be used for most of the problems in this book. When using the calculator, the answer should be estimated first. You should develop this habit of estimating the answer first and practice it whenever you do math. For practice, it is also recommended that every fifth problem be worked out by hand completely before using the calculator.

HINTS ABOUT ESTIMATING

As a check to see whether the correct number has been entered into the calculator and the correct button pushed, you should have an idea what the correct answer would be. You should make this estimate as easy to determine as possible. What you really do is make a very easy math problem that is similar to the actual one you are trying to solve.

You should make the problem so easy that you can do it in your head. To do this, you first need to "round" the numbers off. Usually, the better you round off, the closer you will get to the actual answer. You should round to one-digit numbers. So look at the next digit. If it is 5 or above, round your number up. If it is 4 or less, keep your number as it is. You should round off 55 to 60 and 54 to 50. Round 149 to 100 because the second digit is a 4.

Next, determine the mathematical operation that needs to be performed and perform it. You can "round off" the answer. The result should be close to the actual answer. Remember that this is just the estimate. You have to go back and do the actual calculation. Examples of estimating answers are provided throughout the book.

INTRODUCTION TO MEASUREMENT

There are two parts to any measurement—the number (how many) and the unit (of what). We want to look at the units we use. Two major systems of measurement are used worldwide—English and metric. It is usual to work in one of the two systems at a time. You must be able to recognize the units and the system those units belong to. Let us look at some categories of measurement and what units belong to those categories.

One category has the same units for both measurement systems. That category is time. The units for time include seconds, minutes, hours, and days.

Mass or weight has English units of ounces, pounds, and tons. The metric units for mass or weight include grams and kilograms.

The English units for distance (length) include inches, feet, yards, and miles. The metric units for distance include centimeters, meters, and kilometers.

Area is the size of a flat surface. English units of area include square inches, square feet, and square yards. Metric units for area include square centimeters and square meters.

Volume is the size of the interior of a shape. English units of volume include cubic inches, cubic feet, and cubic yards. Metric units for volume include cubic centimeters and cubic meters.

Units can be of help in telling you if you are doing the problem correctly. If a problem asks how long it took to complete a job and you get an answer of $16.43, you know that you did something wrong because dollars is not a measure of time (how long it took). So if a problem asks for the length of a piece of conduit and you determine the answer is 16 seconds, you know that you made an error. If you are asked to find the area of a window and you determine that it is 12 square feet, then there is a good chance that you worked the problem correctly.

We have to be careful with the units when working math problems. When adding or subtracting, the units must be the same so the answer will have the same unit. If they are not the same, an error will be made working the problem. On the other hand, when multiplying or dividing, the unit of the answer can be different from the units used in the calculation. As an example, a closet measures 3 feet by 4 feet. Notice that each of the measurements has the unit of feet. The floor area of this closet is 12 square feet. Square feet is a different unit than feet, but in this case, it is the correct unit. We found the area by multiplying the length by the width. In doing this, we also multiplied feet by feet and wound up with square feet. We still have to be careful that when we multiply similar quantities, we are using the same units. We should not multiply a length measured in feet by a width measured in inches. They should both be feet or both be inches.

There are times when two different quantities get multiplied or divided. The units are carried through with the numbers. As examples, torque is determined by multiplying a distance (measured in feet) by a force (measured in pounds). The result has the unit of foot-pounds. Speed is determined by dividing distance (measured in miles) by time (measured in hours). The result is speed with the unit of miles per hour. These are correct units.

When working problems throughout this text, pay attention to the units. Working the units properly is just as important as working the numbers correctly.

PPM

Practical Problems in Mathematics

FOR RENEWABLE ENERGY TECHNICIANS

SECTION

Whole Numbers

UNIT 1

Basic Principles of Addition of Whole Numbers

- Study addition of denominate numbers in Section I of the Appendix.

When a number is written, each numeral holds a specific position. Each position has a different value and a different name. Look at the large whole number 193,456,782. The number 2 is in the units position, 8 in the tens position, 7 in the hundreds position, 6 in the thousands position, 5 in the ten thousands position, 4 in the hundred thousands position, 3 in the millions position, 9 in the ten millions position, and 1 in the hundred millions position. The commas group the number in sets of 3 (from the right-hand side of the number). They help one when saying the number. When saying this number, the last three numbers are said as seven hundred eighty-two. Each other group of three numbers is said similarly, with the addition of the word thousand for the next group of three and million for the next group of three (and billion for the next group of three). The large number above is said as one hundred ninety-three million four hundred fifty-six thousand seven hundred eighty-two.

Whole numbers are numbers that have nothing smaller than a unit. There are no fractions of numbers in whole numbers.

Addition is the process of finding the total (sum) of two or more numbers.

When adding whole numbers, it is best to place the numbers in columnar form. To do this, line the units places of each number underneath each other. By lining up the units, all of the other places are also lined up. Some numbers are smaller numbers and do not have the same number of numerals, but that is okay.

EXAMPLE 1: Find the sum of the following:

$$
\begin{array}{r}
11 \\
23 \\
+\,45 \\
\hline
\end{array}
$$

The first step is to estimate the answer so that you have an idea what the answer should be and whether you made a mathematical mistake. Start by rounding the numbers to ones with only one numeral that is different from zero. In our problem all of the numbers will round off to numbers in the tens position.

$$
\begin{array}{r}
11 \\
23 \\
+\,45 \\
\hline
\end{array}
\qquad \rightarrow \qquad
\begin{array}{r}
10 \\
20 \\
+\,50 \\
\hline
\end{array}
$$

Adding these gives 80

So the correct answer should be close to 80. Getting an answer close to 8 is much too small, and an answer close to 800 is too large. Also, if an answer close to 20 or 40 is gotten, you know that a math error has been made, because the answer should be close to 80.

Now do the actual problem

$$
\begin{array}{r}
11 \\
23 \\
+\,45 \\
\hline
\end{array}
$$
Add the units column and place that number under the units column.
$$
\begin{array}{r}
11 \\
23 \\
+\,45 \\
\hline
9 \\
\end{array}
$$
Next add the tens column and place that number under the tens column.
$$
\begin{array}{r}
11 \\
23 \\
+\,45 \\
\hline
79 \\
\end{array}
$$

The answer is 79. That is very close to the estimated answer.

EXAMPLE 2: What is the sum of $674 + 19 + 352 + 8$?

It is easiest to write this problem in columnar form. If the unit numbers are lined up under each other, all of the other columns will be properly aligned.

The first step is to estimate the answer. As we did in the previous example, round the numbers to ones with only one nonzero numeral.

$$
\begin{array}{r}
674 \\
19 \\
352 \\
+\ \ 8 \\
\hline
\end{array}
\qquad \rightarrow \qquad
\begin{array}{r}
700 \\
20 \\
400 \\
+\ \ 10 \\
\hline
\end{array}
$$

Adding the columns gives 1,130

So our correct answer sholud be close to 1,130.

Now do the actual problem.

| $\begin{array}{r}674\\19\\352\\+\ \ 8\\\hline\end{array}$ First add the units column. The total is 23. Write the 3 under the units column and put a 2 above the tens column. This is called carrying the 2. | $\begin{array}{r}\overset{2}{6}74\\19\\352\\+\ \ 8\\\hline 3\end{array}$ Now add the tens column including the 2. If no number is in the column, do not add anything. The total for our tens column is 15. Write a 5 under the tens column and carry the 1. | $\begin{array}{r}\overset{1\,2}{6}74\\19\\352\\+\ \ 8\\\hline 53\end{array}$ Add the hundreds column. This gives a total of 10. Since no thousands column exists, we will create one. Write 10 with the 0 in the hundreds column. | $\begin{array}{r}\overset{1\,2}{6}74\\19\\352\\+\ \ 8\\\hline 1,053\end{array}$ |

The total is 1,053. This is close to our estimate of 1,130.

Very often the quantity is made up of two parts. Those parts are "how much" and "of what." The "how much" is called the number or the size or the quantity. The "of what" is called the unit (the same name as the last position in a number, but a totally different meaning). When adding numbers that have units, you must make sure they have the same units. The sum (answer) will have the same unit.

Practical Problems:

• Apply the principles of addition of whole numbers to the problems in this unit.

In problems 1–12, add the numbers.

| 1. | $\begin{array}{r}415\\+\ 523\\\hline\end{array}$ | 2. | $\begin{array}{r}3,115\\1,050\\+\ 2,728\\\hline\end{array}$ | 3. | $\begin{array}{r}6,024\\78\\4,386\\+\ \ 517\\\hline\end{array}$ |

4.   127 gallons   5.   4,344 feet   6.   777 meters
      88 gallons   5,271 feet   68 meters
  + 314 gallons   63 feet   2,458 meters
                 +   403 feet   40 meters
                                      + 8,239 meters

7. 647 + 585 =   _____

8. 5,214 + 1,177 + 3,418 =   _____

9. 45 + 2,208 + 579 + 4,166 =   _____

10. 50 hours + 117 hours + 91 hours =   _____

11. 871 miles + 3,406 miles + 1,985 miles + 1,630 miles =   _____

12. 23,135 MW + 883 MW + 4,729 MW + 644 MW + 5,009 MW =   _____

13. Three wind generators are in a straight line as shown. What is the
    distance from the first generator to the third generator?   _____

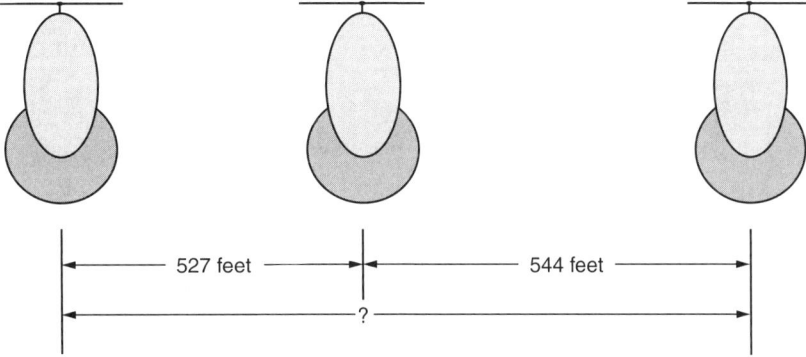

14. Piping needed to move water from a solar heat collector to a storage tank is shown. What is the total length of straight pipe for this system? _____

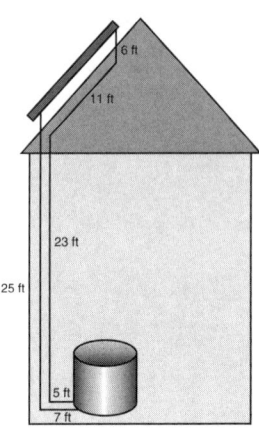

15. Technician Makena worked the following hours one week: Monday – 9 hours, Tuesday – 8 hours, Wednesday – 11 hours, Thursday – 8 hours, Friday – 11 hours, and Saturday – 6 hours. How many hours did Makena work this week? _____

16. Solar data is being collected for the town of Sunny City. The hours of actual sunshine during one week were: Monday – 9 hours, Tuesday – 7 hours, Wednesday – 1 hour, Thursday – 10 hours, Friday – 10 hours, Saturday – 3 hours, and Sunday – 6 hours. What is the total sunshine time for that week? _____

17. Two low-head dams are built on each of two streams on a farm property. The vertical drops from the dams are shown. What is the total vertical drop from these dams? _____

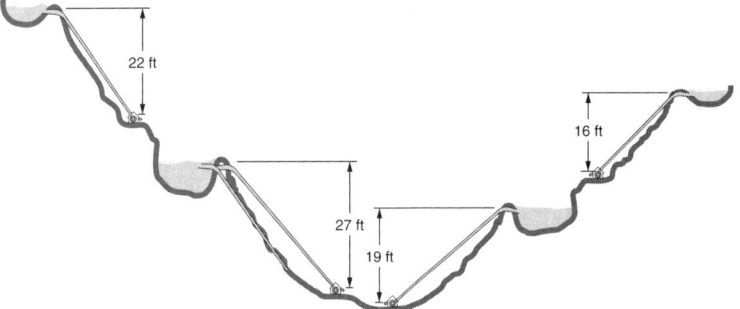

18. Nell's New Energy Wind Farm has 8 large wind generators on it. During one month, these generators produced 38 MWh, 35 MWh, 43 MWh, 40 MWh, 38 MWh, 39 MWh, 39 MWh, and 37 MWh. How many MWh were produced on Nell's Wind Farm that month? _____

19. The Personal Power Company installs and services small, individual wind generators. Technician James drove the following distances making calls one week: 45 miles, 27 miles, 32 miles, 25 miles, and 39 miles. How far did James drive that week? _____

20. Four technicians make service calls for the Suns Up Solar Systems Company. During a slow week, technician Brett made 7 service calls, Parker had 9 calls, Carson had 12 calls, and Hayden had 13 calls. How many service calls were made that week by these technicians? _____

21. Three streams on a property were measured and have 1-minute flows as shown. What is the total stream flow on this property? _____

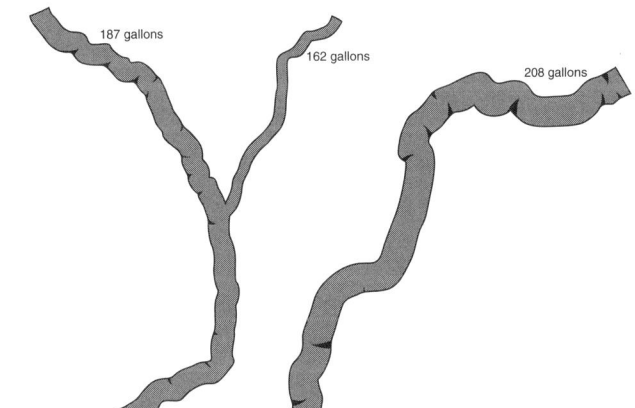

22. Four geothermal wells need to be filled with grout. The weights of the grout put into the wells were 206 pounds, 198 pounds, 248 pounds, and 197 pounds. What is the total weight of the grout used? _____

23. Three wells are drilled for a geothermal home heating system. The well depths are 170 ft, 150 ft, and 180 ft. What is the total well depth dug for this system? _____

24. The Personal Power Company has 5 service trucks. Each truck carries #8 cable in them. Truck 1 has 352 ft of cable, truck 2 has 117 ft, truck 3 has 327 ft, truck 4 has 289 ft, and truck 5 has 88 ft. What is the total amount of cable in the trucks?   _____

25. Conduits carry the energy from wind generators to the distribution panel for the wind farm as shown. The conduits run directly from the generator to the panel. Find the total length of conduit needed.   _____

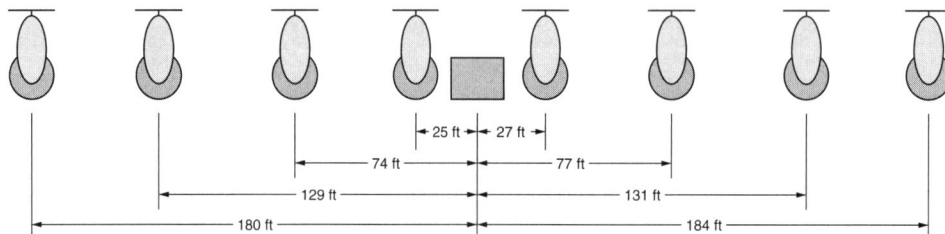

UNIT 2

Subtraction of Whole Numbers

Basic Principles of Subtraction of Whole Numbers

• Study subtraction of denominate numbers in Section I of the Appendix.

Subtraction is the process of finding the difference between two numbers.

When subtracting, place the smaller number underneath the larger one and line up the units columns. The other columns will then be properly aligned. Begin with the units column and subtract the bottom number from the top number. If there is no number on the bottom, the top number is brought down to the answer.

If the numbers have measurement units associated with them, they must have the same units before they can be subtracted.

EXAMPLE 1:   Subtract 431 from 858.

Estimate the answer.

$$
\begin{array}{r}
858 \\
-\ 431 \\
\hline
\end{array}
\quad\rightarrow\quad
\begin{array}{r}
900 \\
-\ 400 \\
\hline
\end{array}
$$

Subtracting gives 500

Now do the actual problem.

$$
\begin{array}{r}
858 \\
-\ 431 \\
\hline
427
\end{array}
$$

EXAMPLE 2: Find the difference between 1,246 and 689.

Estimating

$$
\begin{array}{r}
1{,}246 \\
-\ 689 \\
\hline
\end{array}
\qquad \rightarrow \qquad
\begin{array}{r}
1{,}000 \\
-\ 700 \\
\hline
300
\end{array}
$$

Doing the actual problem:

$$
\begin{array}{r}
{}^{\;3\;16}\\
1{,}2\overset{}{4}\overset{}{6} \\
-\ 689 \\
\hline
7
\end{array}
$$
It is not possible to subtract 9 from 6, so "borrow" 1 from the next column, making the 4 a 3. The 1 becomes a 10 and is added to the 6, making it a 16. 9 can now be subtracted from the 16.

$$
\begin{array}{r}
{}^{1\;13\;16}\\
1{,}246 \\
-689 \\
\hline
57
\end{array}
$$
Now the 8 cannot be subtracted from the 3, so we "borrow" 1 from the next column, making the 2 a 1. The borrowed 1 becomes a 10 and makes the 3 a 13. 8 can be subtracted from the 13.

$$
\begin{array}{r}
{}^{1\;13\;16}\\
1{,}246 \\
-689 \\
\hline
557
\end{array}
$$
Finally, the 6 can be subtracted from the 11, giving the final answer.

The 557 is close to the estimated 300.

Practical Problems:

- Apply the principles of subtraction of whole numbers to the problems in this unit.

In problems 1–12, subtract the numbers.

1.
$$
\begin{array}{r}
648 \\
-\ 212 \\
\hline
\end{array}
$$

2.
$$
\begin{array}{r}
5{,}293 \\
-\ 1{,}387 \\
\hline
\end{array}
$$

3.
$$
\begin{array}{r}
7{,}104 \\
-\ 856 \\
\hline
\end{array}
$$

4.
$$
\begin{array}{r}
774 \text{ meters} \\
-\ 351 \text{ meters} \\
\hline
\end{array}
$$

5.
$$
\begin{array}{r}
12{,}462 \text{ MW} \\
-\ 9{,}917 \text{ MW} \\
\hline
\end{array}
$$

6.
$$
\begin{array}{r}
6{,}257 \text{ hours} \\
-\ 3{,}348 \text{ hours} \\
\hline
\end{array}
$$

7. $857 - 643 =$ _____

8. $2{,}416 - 2{,}388 =$ _____

9. $8{,}962 - 4{,}473 =$ _____

10. 533 miles $-$ 212 miles $=$ _____

11. 9,378 feet $-$ 3,547 feet $=$ _____

12. 27,508 gallons $-$ 13,129 gallons $=$ _____

13. A wind generator tower is to be 127 feet high when finished. The tower is under construction and 59 feet have been erected. How many feet of the tower still need to be added? _____

14. A photovoltaic (PV) system is to be installed. It will take 250 man-hours to do the installation. 77 man-hours have been worked so far. How many man-hours are needed to finish the job? _____

15. Technician Makena started the day with 1,325 ft of #4/0 wire on her truck. By noon, she had used 810 ft of wire. How many feet of wire does Makena have on her truck at the start of the afternoon? _____

16. Tubing used to connect the solar collector with the storage tank has a straight section that is 37 feet long. Technician Hayden has one piece of straight tubing that is 14 feet long. How much straight tubing does Hayden need to finish this job? _____

17. A house uses 3,517 kWh of electrical power one month. A wind generator produced 2,825 kWh of power for that same month. The rest of the energy used must be purchased from the power company. How many kWh must be purchased from the power company for this month? _____

18. A dam has its water level at the elevation above sea level shown. A place to locate the low-head water turbine is also shown. What is the height that the water falls to the turbine?
_____

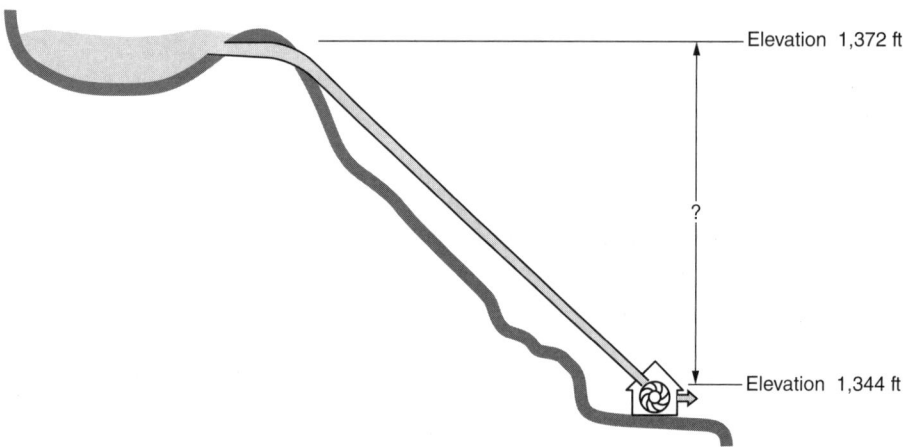

Elevation 1,372 ft

?

Elevation 1,344 ft

19. How far away from the ground is the tip of the wind turbine blade at its lowest point?
_____

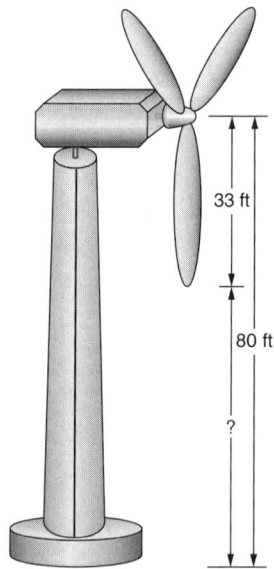

33 ft

80 ft

?

20. A solar collector has water flow through it as shown. The water is heated how many degrees in the collector?

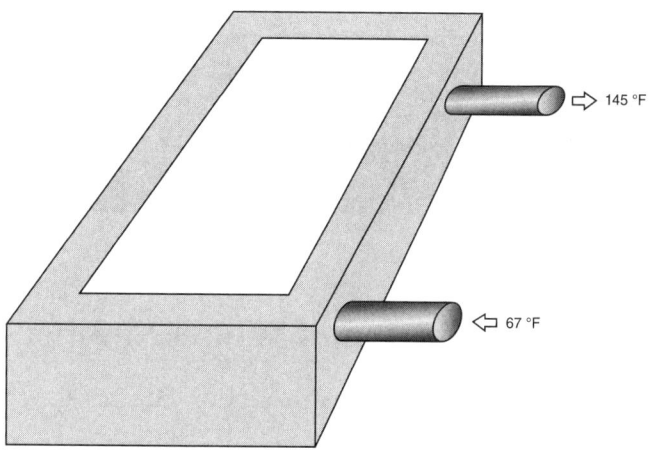

145 °F

67 °F

21. 117 square feet of PV collector is needed for an installation. Technician Parker has 45 square feet in stock. How many square feet of PV collector must Parker order to complete this installation?

22. 100 ft of #6 cable weighs 19 pounds. 100 ft of #4/0 cable weighs 86 pounds. How much heavier is 100 ft of #4/0 cable than 100 ft of #6 cable?

23. A 1,500 kW wind turbine has blades measuring 159 ft in length. A 225 kW wind turbine has 33 ft blades. How much longer is the 1,500 kW blade than the 225 kW blade?

24. During one 43-hour work week, technician Carson spent 9 hours driving to wind generator locations after spending 6 hours at the warehouse getting supplies needed to service the generators. The rest of the time Carson spent servicing the generators. How many hours did Carson spend servicing the wind generators?

25. Three wells are being drilled for a geothermal system. The total well depth needed is 1,800 feet. One well has a depth of 575 feet. The second well is 610 feet deep. How deep must the third well be?

UNIT 3

Multiplication of Whole Numbers

Basic Principles of Multiplication of Whole Numbers

- Study multiplication of denominate numbers in Section I of the Appendix.

When multiplying two numbers, place one under the other, lining up their units column. Next, multiply the entire first number by the units number of the second number. Next multiply the entire first number by the tens number of the second, lining up the first number with the tens column of the previous number. Continue until the first number has been multiplied by each of the numbers in the second number, lining the products up one place to the left each time. Finally, add each of the columns to get the final product.

Unlike addition or subtraction, the numbers do not have to and usually do not have the same units. But the units do get multiplied together.

EXAMPLE 1: A wind turbine needs 215 ft of electrical cable to connect the generator to the distribution panel at the base of the turbine. How much cable is needed for 13 turbines?

This is a multiplication problem. We will multiply 215 ft $\times$ 13. First we estimate the answer.

$$200 \text{ ft} \times 10 = 2{,}000 \text{ ft}$$

Both numbers were rounded down, so this estimate will be lower than the actual answer, but this gives us an idea as to what the answer should be. Now let us work the problem.

$$
\begin{array}{r}
215 \text{ ft} \\
\times\ 13 \\
\hline
\end{array}
$$

14

First, multiply 215 by 3. When multiplying, we may need to carry numbers—in our case, $3 \times 5 = 15$, so we write the 5 and carry the 1. Multiply the 3 by 1, giving 3, but then add the carried 1 to get 4. Multiply the 3 by 2, giving 6.

$$
\begin{array}{r}
^{1} \\
215 \text{ ft} \\
\times\ 13 \\
\hline
645
\end{array}
$$

Next, multiply 215 by 1, but place the 5 directly under the 1 in the 13.

$$
\begin{array}{r}
215 \text{ ft} \\
\times\ 13 \\
\hline
645 \\
215
\end{array}
$$

Now add the columns of the products to arrive at our answer.

$$
\begin{array}{r}
215 \text{ ft} \\
\times\ 13 \\
\hline
645 \\
215 \\
\hline
2{,}795 \text{ ft}
\end{array}
$$
     The units are brought down to the answer.

If the numbers being multiplied have different units, the answer has the product of the units. As examples: ft $\times$ ft = sq ft, and ft $\times$ lbs = ft-lbs.

If three numbers are multiplied together, multiply two of them together. Then take that product and multiply it by the third number. Continue in this manner if four or more numbers are to be multiplied.

EXAMPLE 2: During an installation project, 2 small gasoline generators were each run 6 hours on 3 different days. What is the total number of hours that the gasoline generators were run on this installation?

We need to multiply 2×6 hr $\times 3$. This is a relatively easy problem, so we do not need to estimate the answer. We would not expect the answer to be very large.

First, multiply

$$\begin{array}{r} 2 \\ \times\ 6\ \text{hr} \\ \hline 12\ \text{hr} \end{array}$$

Now multiply the 12 hr by 3.

$$\begin{array}{r} 12\ \text{hr} \\ \times\ 3 \\ \hline 36\ \text{hr} \end{array}$$

Practical Problems

• Apply the principles of multiplication of whole numbers to the problems in this unit.

In problems 1–12, multiply the numbers.

1. $\begin{array}{r} 13 \\ \times\ 32 \\ \hline \end{array}$

2. $\begin{array}{r} 476 \\ \times\ 57 \\ \hline \end{array}$

3. $\begin{array}{r} 2{,}089 \\ \times\ 91 \\ \hline \end{array}$

4. $\begin{array}{r} 844\ \text{gallons} \\ \times\ 15 \\ \hline \end{array}$

5. $\begin{array}{r} 16\ \text{feet} \\ \times 12\ \text{feet} \\ \hline \end{array}$

6. $\begin{array}{r} 2{,}407\ \text{MW} \\ \times\ 24\ \text{hours} \\ \hline \end{array}$

7. $92 \times 77 =$ _____

8. $4{,}005 \times 68 =$ _____

9. $357 \times 646 =$ _____

10. 519 meters $\times 26 =$ _____

11. 43 centimeters $\times 52$ centimeters $=$ _____

12. 744 pounds $\times 219 =$ _____

13. One blade of a 1,500 kW wind turbine weighs 24,000 pounds. What is the total weight of 15 blades? _____

14. A PV cell has the surface area shown. What is the surface area of an array of 24 cells?   _____

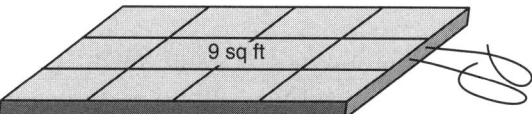

15. A geothermal well needs 172 feet of tubing. How many feet of tubing would be needed for 18 wells?   _____

16. The warehouse has 75 sections of electrical conduit. Each section is 12 feet long. What is the total length of conduit in the warehouse?   _____

17. Suzanne, the manager of the 'Big Wind' Wind Farm has determined it takes 28 hours to overhaul a 1.2 MW turbine. How many hours does Suzanne need to schedule to overhaul 17 turbines?   _____

18. Nell was hired to assist a crew installing a wind turbine farm. Her main task was to drive to town and back getting supplies for the crew. During one week, Nell made 16 trips to town and back. Each trip was a total of 53 miles. How far did Nell travel that week?   _____

19. Five identical wind turbines are planned to be built equally spaced in a straight line. What is the distance from the first tower to the last tower?   _____

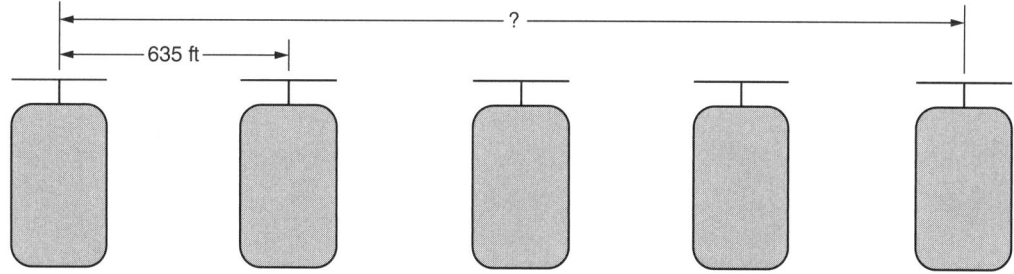

20. A 4 in diameter geothermal well 500 ft deep requires 45 cubic feet of grout. During one August, the Hot Stuff Geothermal System Company drilled 18 wells that were 500 ft deep. How many cubic feet of grout were needed for the wells drilled that August?

    _____

21. 112 pounds of slurry produce 1 cubic foot of grout for the geothermal wells in the previous problem. What is the weight of the slurry needed for the wells drilled that August?

    _____

22. A large PV system was installed on a university campus. 23 technicians worked on the project 6 days each week. They worked 7 weeks working 11 hours each day. What was the total number of hours worked on this project?

    _____

23. A low-head hydro turbine uses a water flow of 35 gallons each second. What is the number of gallons that flow through the turbine in one day?

    _____

24. One style of wind turbine blade is attached to the hub by 16 bolts. Each turbine has 3 blades. How many bolts are used to attach turbine blades to the 14 turbines on one wind farm?

    _____

25. A wind generator averages 225 kW each hour. What would be this generator's output for an average week in kWh?

    _____

UNIT 4

Division of Whole Numbers

Basic Principles of Division of Whole Numbers

• Study division of denominate numbers in Section 1 of the Appendix.

Division is one of the most difficult of the mathematical operations. But it can be done correctly if care is taken to follow the correct procedure each time.

Each of the numbers in a division problem has a special name.

$$\overset{\text{QUOTIENT}}{\text{DIVISOR }\overline{)\text{DIVIDEND}}}$$

The division problem could have also been written as DIVIDEND ÷ DIVISOR = QUOTIENT. The word "by" is sometimes used in place of the ÷ sign.

To solve a division problem, first write down the problem.

EXAMPLE 1:   Divide 291,772 by 62.

We first estimate the answer by changing the problem to: Divide 300,000 by 60.

This is written as $60\overline{)300,000}$

Divide 60 into 3. It cannot be done. So then try dividing 60 into 30. This also cannot be done. Next try dividing 60 into 300. This gives exactly 5. We put the 5 in the quotient directly above the 0 that made the 300.

$$\overset{5}{60\overline{)300,000}}$$

There is nothing else to divide the 60 into. There are only 0s. Add 0s to the quotient above each of the 0s in the dividend that have not been used yet.

$$\begin{array}{r} 5{,}000 \\ 60\,\overline{)300{,}000} \end{array}$$

The correct answer should be close to 5,000.

Let us now solve the problem.

$$62\,\overline{)291{,}772}$$

Divide 62 into 2. It cannot be done. Divide 62 into 29. It also cannot be done. Divide 62 into 291. This gives 4. Write the 4 in the quotient above the 1 position in the dividend. Multiply 4×62 and put 248 beneath 291.

$$\begin{array}{r} 4, \\ 62\,\overline{)291{,}772} \\ \underline{248} \end{array}$$

Now subtract 248 from 291. This gives 43. Divide 62 into 43. It cannot be done. (If it could, you made a mistake in the 4. It should be a 5 and we would be subtracting a larger number from the 291.) We cannot divide 62 into 43, so we bring down the first 7, making 437.

$$\begin{array}{r} 4, \\ 62\,\overline{)291{,}772} \\ \underline{248} \\ 43\ 7 \end{array}$$

Dividing 62 into 437 gives 7. Put that above the 7 and multiply 7×62. Write the product below the 437 and subtract.

$$\begin{array}{r} 4,7 \\ 62\,\overline{)291{,}772} \\ \underline{248} \\ 43\ 7 \\ \underline{43\ 4} \\ 37 \end{array}$$

The second 7 is brought down, making 37. Divide 62 into 37. It cannot be done. Now a 0 is placed in the quotient above the second 7. Then the 2 is brought down, making 372.

$$
\begin{array}{r}
4,70 \\
62\overline{)291,772} \\
248 \\
\overline{437} \\
434 \\
\overline{372} \\
\end{array}
$$

Divide 62 into 372. This gives 6. Place the 6 in the quotient. Multiply 6 $\times$ 62 and place that under 372 and subtract. This gives 0, so the problem is complete.

$$
\begin{array}{r}
4,706 \\
62\overline{)291,772} \\
248 \\
\overline{437} \\
434 \\
\overline{372} \\
372 \\
\overline{0} \\
\end{array}
$$

The answer to divide 291,772 by 62 is 4,706, which is close to our estimated 5,000.

There are times when the last subtraction results in a number other than 0.

EXAMPLE 2: Divide 7 by 3.

$$
3\overline{)7}
$$

Dividing 3 into 7 gives 2.

$$
\begin{array}{r}
2 \\
3\overline{)7} \\
6 \\
\end{array}
$$

Subtracting gives 1. 3 does not divide into 1 and there are no other numbers. The 1 is known as a remainder. So 7 divided by 3 is 2 with a remainder of 1.

As with multiplication, the numbers do *not* have to (and usually do not) have the same units. But the units must be treated just as the numbers were. For example: miles ÷ hour = miles per hour, sq ft ÷ ft = ft.

Practical Problems:

• Apply the principles of division of whole numbers to the problems in this unit.

In problems 1–12, divide the numbers.

1. $3\overline{)639}$

2. $47\overline{)1,128}$

3. $714\overline{)239,190}$

4. $4\overline{)804}$ pounds

5. $92\overline{)37,720}$ feet

6. $506\overline{)356,224}$ MW

7. $656 \div 8 =$ _____

8. $1,386 \div 18 =$ _____

9. $996,839 \div 2,153 =$ _____

10. $123,284$ m $\div 476 =$ _____

11. $277,692$ sq in $\div 634 =$ _____

12. 648 miles $\div 8$ hours $=$ _____

13. To make connection with the electrical grid, a wind farm must run a line 3,145 ft. Poles to carry the line must be placed every 185 ft. How many poles are needed?   _____

14. A solar collector is planned for a hot water heating system for a Laundromat. One collector unit is shown. How many collector units are needed for the system?   _____

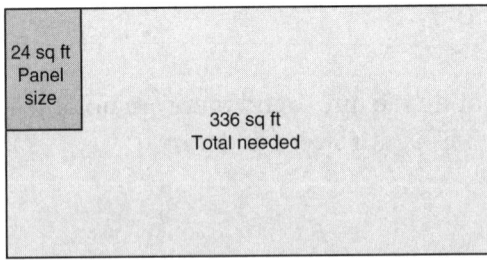

24 sq ft Panel size

336 sq ft Total needed

15. A low-head hydro turbine generator produces 5 kW with a flow of 35 gallons per second. What flow is needed to produce each kW? _____

16. A project to install wind turbines is being planned. Project manager Suzanne estimates the job will take 126 total hours. If she asks the installers to work 9 hours each, how many installers should Suzanne assign to the job to complete it in one day? _____

17. A small wind turbine has a capacity of 225 MW. How many turbines will be needed for a capacity of 5,175 MW? _____

18. PV collectors are stored on pallets in a warehouse as shown. How many collectors can be stored on one pallet? _____

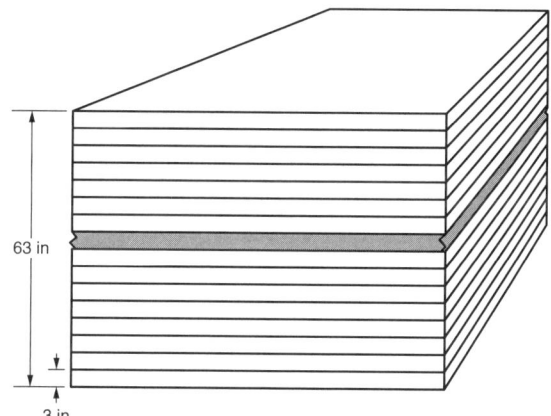

63 in

3 in

19. A geothermal well needs to be drilled. It is to be 720 feet deep. On the average, the drilling rig can drill 45 feet each hour it runs. How long will it take to drill this well? _____

20. A direct solar hot water system has a storage tank of 280 gallons. It is desired to have the water be recycled 6 times each hour. What must the flow rate through the collector be each minute to cause 6 turnovers each hour? _____

21. Chain-link fence surrounds each wind turbine tower as shown. How many tower sides can be fenced with 500 feet of chain-link fencing? How much fencing is left over?   _____

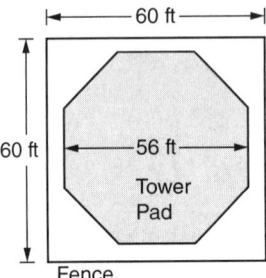

22. A stream is determined to have 133,200 gallons flow past a point in one hour. What is the flow rate of the stream in gallons per second?   _____

23. A geothermal heating system has a well that is 525 feet deep and supplies a 3-ton cooling system. What is the foot depth per ton of cooling for this well?   _____

24. The output of a PV unit is 240 watts. What is the output per sq ft for this unit?   _____

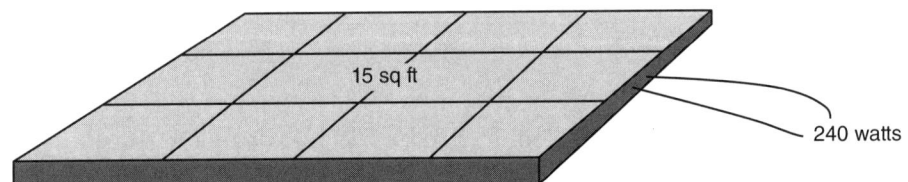

25. A geothermal heating/cooling system has 3 ground heat exchange branches. The total flow splits into 3 and goes through one of the branches. If the total system has a flow of 540 gallons in an hour, what is the flow through one branch?   _____

UNIT 5

Combined Operations with Whole Numbers

Basic Principles of Combined Operations with Whole Numbers

- Review and apply the principles of addition, subtraction, multiplication, and division of whole numbers to the problems in this unit.

The word problems in this unit are more like real-life problems than the ones in the preceding units. This is because you are not told which mathematical operation you need to perform in order to get the answer. So part of the problem solution is to determine whether to add, subtract, multiply, or divide. You also have to determine which numbers you need to use. There are times where numbers are given which are not used to determine the answer. You just have to ignore those numbers. Once you have chosen which numbers to use and which operation to perform on those numbers, next you have to correctly perform the operation or operations (sometimes you need to perform more than one operation). Finally, you need to ask yourself, "Are there any units associated with the numbers I am using that must be put into the answer?"

There are often hints in the way the problem is worded that give a clue as to which operation to perform. These hints will not work in every case, but they may help determine the correct operation.

1. If the problem asks for the **total**, it is either an addition or a multiplication problem.

2. If the problem asks you to find the **difference**, the problem is a subtraction problem.

3. If the problem asks you to determine something **for each**, it usually is a division problem.

You should develop a series of hints that work for you. One way of doing that is to practice by working a lot of problems.

EXAMPLE: A wind farm experiences wind strong enough to produce the maximum generation for two hours. The farm has three 225 kW generators and five 450 kW generators. What was the total generation during these two hours?

This problem is solved using multiple operations. To estimate the answer, multiply 3×200 kW = 600 kW. Then multiply 5×500 kW = 2,500 kW. Add these together: 600 kW + 2,500 kW = 3,100 kW. Multiply this total by 2 hours, 3,100 kW $\times$ 2 hours = 6,200 kWh.

To solve the exact problem, first, multiply 3×225 kW = 675 kW. Next, multiply 5×450 kW = 2,250 kW. Next we add those two numbers together: 675 kW + 2,250 kW = 2,925 kW. Finally, we need to multiply 2,925 kW $\times$ 2 hours = 5,850 kWh. This is close to our estimate.

Practical Problems:

1. $\begin{array}{r} 371 \\ 520 \\ 188 \\ + 643 \\ \hline \end{array}$

2. $\begin{array}{r} 2,705 \\ 88 \\ 3,461 \\ + 947 \\ \hline \end{array}$

3. $\begin{array}{r} 7,625 \\ - 2,311 \\ \hline \end{array}$

4. $\begin{array}{r} 18,206 \\ - 4,319 \\ \hline \end{array}$

5. $\begin{array}{r} 692 \\ \times \ \ 47 \\ \hline \end{array}$

6. $\begin{array}{r} 5,837 \\ \times \ \ \ 207 \\ \hline \end{array}$

7. $45\overline{)16,515}$

8. $703\overline{)4,251,744}$

9. $3,244 + 679 =$ _____

10. $6,482 - 517 =$ _____

11. $1,752 \times 639 =$ _____

12. $836,976 \div 371 =$ _____

13. A wind farm has 7,436 MW of generation. A 686 MW generator is added to the farm. What is the new generation capacity of the farm? _____

14. A PV array produces 6,480 watts. A unit producing 360 watts is defective and is removed from the array. What is the new wattage? _____

15. A geothermal system consists of 2 trenches, shown with heat exchange copper tubing out and back in each trench. What is the total length of parallel tubing in the trenches for this system? _____

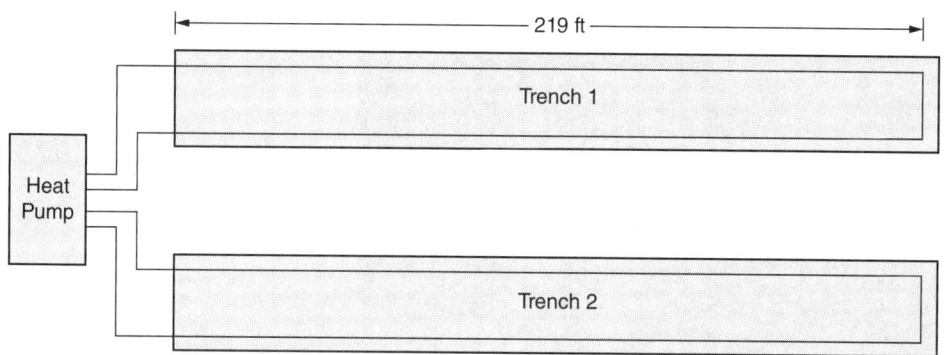

16. A major solar array was installed. 12 workers worked a total of 4,536 hours. How many hours did each worker spend on the installation? _____

17. A low-head hydro system is being fed from a new dam. What will be the new height of head for the turbine? _____

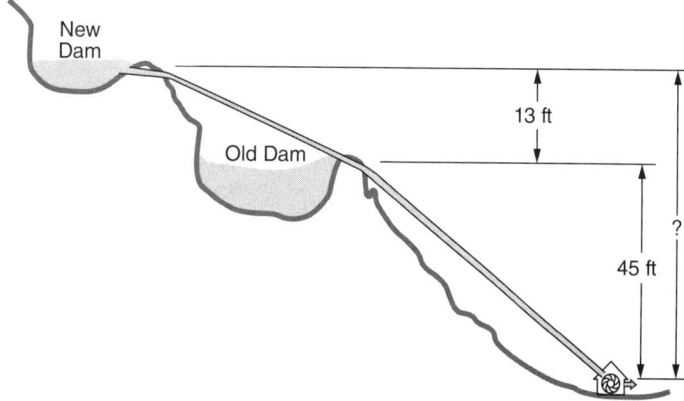

18. A geothermal system runs fluid into and out of a well. What is the temperature change of the fluid? _____

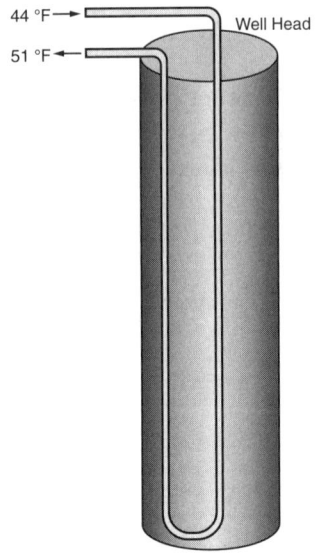

19. Each home in a housing development was built with a PV array that can generate 1,080 W in an hour. There are 57 homes in the development. How much generation capability does this development have? _____

20. A narrow strip of mountaintop land 2 miles long (10,560 feet) is being considered for a wind farm. The center of each turbine must be at least 660 feet from the center of the next one. What is the maximum number of turbines that could be put on this land? (Hint: Put one at the beginning and then each of the additional ones is 660 feet apart. So your answer should include 1 more.) _____

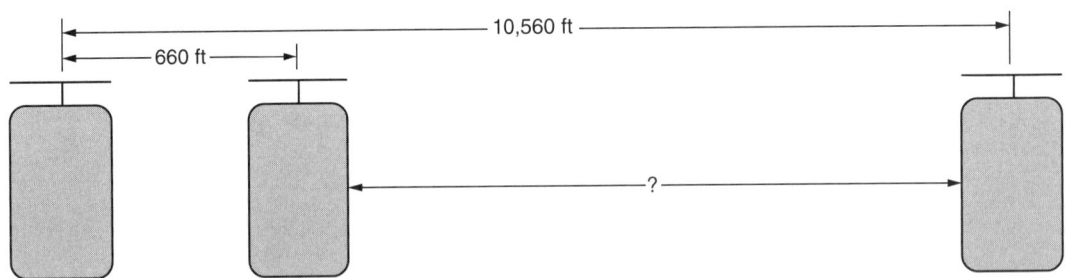

21. One house has 450 sq ft of solar collector on its roof and 180 sq ft of solar collector on the garage roof. What is the total area of solar collector?   _____

22. A PV array needs to average 240 kWh each month to pay for itself. When operating for an entire sunny day, it produces 16 kWh. How many sunny days must the PV array average each month to pay for itself?   _____

23. One bag of grout, when mixed, will fill a volume of 5 cu ft in a geothermal well. 12 bags of grout are needed to fill a well. What is the volume of the well dug?   _____

24. There are 36 turbines on Nell's wind farm. It takes 23 gallons of paint to paint the tower of a wind turbine. The painter can apply 4 gallons of paint each hour. How long does the painter need to paint all of the turbines on Nell's farm?   _____

25. A repair job on a turbine for a low-head generator is projected to be 184 hours. Technician Brett has worked 72 hours on the repair so far. How many more 8-hour days should Brett expect to work on this job?   _____

SECTION 2

Common Fractions

UNIT 6

Addition of Common Fractions

Basic Principles of Addition of Common Fractions

- Study addition of denominate numbers in Section I of the Appendix.

It is best to work problems involving common fractions longhand. Some common fractions have exact decimal fraction equivalents. These can be converted and solved using a calculator. However, there are a number of common fractions that do not have exact decimal equivalents. These can only be accurately solved by working them out longhand. So, let us learn how to solve problems.

Each fraction is made up of two numbers:

$$\frac{\text{NUMERATOR}}{\text{DENOMINATOR}}$$

If there is a whole number along with the fraction, the combination is called a mixed number.

When adding fractions, they can be arranged in a vertical column or a linear (horizontal) form, just as whole numbers were. In either case, the same procedure must be followed. Fractions cannot be added unless they have the same denominators.

If the fractions have the same denominators, add the numerators and place the total over the denominator.

EXAMPLE 1: Add $\frac{1}{7}$ and $\frac{4}{7}$.

$$\frac{1}{7} + \frac{4}{7} = \frac{1 + 4}{7} = \frac{5}{7}$$

The same process applies when using the vertical form.

EXAMPLE 2: What is the sum of $\frac{3}{5}$ and $\frac{1}{5}$?

$$\frac{3}{5}$$
$$+\frac{1}{5}$$
$$\frac{3+1}{5} = \frac{4}{5}$$

The answers can also be estimated. The fractions should be rounded to whole numbers for the estimation. When rounding, if the fraction is equal to or greater than $\frac{1}{2}$, any units number part of the mixed number is increased by 1. If there is no mixed number (just the fraction), make a whole number 1. If the fraction is less than $\frac{1}{2}$, leave the units number unchanged or make a whole number 0. The estimate is then made in the same way as was discussed in whole numbers.

If the fractions to be added do not have the same denominator, equivalent fractions with the same denominators must be created, and those are added together. This is called finding a common denominator. The way common denominators are found is to take the largest denominator and then make multiples of it (1 × the denominator, then 2 × the denominator, and so on), checking each time to see if all of the other denominators divide evenly into that number.

EXAMPLE 3: Find the sum of $\frac{1}{6}$, $\frac{1}{3}$, and $\frac{3}{8}$.

The denominators are different, so we need to find a common denominator. Start with 8, the largest denominator. Neither 6 nor 3 divide evenly into 8, so try 2 × 8 = 16. Again, neither 6 nor 3 divide evenly into 16. Next try 3 × 8 = 24. Both 6 and 3 divide evenly into 24, so 24 is a common denominator. We next make equivalent fractions with 24 as the denominator. We make those fractions by multiplying both numerator and denominator by the same number. That number that is multiplied is the number that will make the denominator the common denominator. The number may be different for each fraction.

$$\frac{1}{6} + \frac{1}{3} + \frac{3}{8} = \frac{1 \times 4}{6 \times 4} + \frac{1 \times 8}{3 \times 8} + \frac{3 \times 3}{8 \times 3} = \frac{4}{24} + \frac{8}{24} + \frac{9}{24} = \frac{4+8+9}{24} = \frac{21}{24}$$

The last step of a problem with fractions is to see if the fraction can be reduced. This means seeing if there is a number that will divide evenly into both the numerator and the denominator.

In the answer above, both 21 and 24 can be divided evenly by 3. Divide both by the common term and drop that common term. The new fraction is equivalent to the old one.

$$\frac{21}{24} = \frac{21 \div 3}{24 \div 3} = \frac{7}{8}$$

This is known as reducing the answer to lowest terms.

There are times when the numerator in the answer is larger than the denominator.

EXAMPLE 4: Add $\frac{5}{9} + \frac{8}{9} + \frac{1}{9} + \frac{7}{9}$.

$$\frac{5}{9} + \frac{8}{9} + \frac{1}{9} + \frac{7}{9} = \frac{5 + 8 + 1 + 7}{9} = \frac{21}{9}$$

There are two multiples of 9 in the numerator, so take them out and make them an equivalent whole number. Then see if the remaining fraction can be reduced. In this case, it can.

$$\frac{21}{9} = 2\frac{3}{9} = 2\frac{1}{3}$$

When adding mixed numbers (or whole numbers with mixed numbers or mixed numbers with fractions), properly add the fractions, then add the whole numbers. Finally, try to reduce the fraction and adjust the whole number part of the answer.

EXAMPLE 5: Find the sum of $2\frac{3}{4}$ and $4\frac{7}{8}$.

The common denominator is 8.

$$\frac{3}{4} + \frac{7}{8} = \frac{3 \times 2}{4 \times 2} + \frac{7 \times 1}{8 \times 1} = \frac{6}{8} + \frac{7}{8} = \frac{6 + 7}{8} = \frac{13}{8} = 1\frac{5}{8}$$

Then the whole numbers are included.

$$2 + 4 + 1\frac{5}{8} = 7\frac{5}{8}$$

This same example could be solved a different way. The mixed numbers can be converted to all fractions (called improper fractions). The improper fractions are then added following the rules for addition of fractions. The total fraction is then reduced to lowest terms.

$$2\frac{3}{4} + 4\frac{7}{8} = \frac{8+3}{4} + \frac{32+7}{8} = \frac{11}{4} + \frac{39}{8} = \frac{11 \times 2}{4 \times 2} + \frac{39 \times 1}{8 \times 1}$$

$$= \frac{22}{8} + \frac{39}{8} = \frac{22+39}{8} = \frac{61}{8} = 7\frac{5}{8}$$

The answer is the same solving the problem using either method. Either way is correct.

Remember and use the following guidelines when adding common fractions:

- Find the common denominator.
- Make equivalent fractions using the common denominator.
- Reduce the fraction in the answer to the lowest terms.

Practical Problems:

- Apply the principles of addition of common fractions to the problems in this unit.

In problems 1–12, add the numbers.

1. $\frac{2}{9}$

    $+ \frac{5}{9}$

2. $\frac{1}{4}$

    $+ \frac{3}{8}$

3. $2\frac{1}{6}$

    $+ 1\frac{1}{3}$

4. $\frac{5}{7}$ feet

    $+ \frac{2}{3}$ feet

5.
$$\frac{1}{5} \text{ hour}$$

$$\frac{3}{4} \text{ hour}$$

$$+ \frac{3}{10} \text{ hour}$$
_____

6.
$$4\frac{1}{4} \text{ pounds}$$

$$3\frac{5}{8} \text{ pounds}$$

$$+ 2\frac{5}{6} \text{ pounds}$$
_____

7. $\dfrac{4}{7} + \dfrac{5}{7} =$ _____

8. $\dfrac{3}{8} + \dfrac{7}{8} + \dfrac{5}{9} =$ _____

9. $3\dfrac{3}{11} + 2\dfrac{1}{2} =$ _____

10. $7\dfrac{2}{3} \text{ sq ft} + 2\dfrac{1}{4} \text{ sq ft} =$ _____

11. $\dfrac{4}{5} \text{ meter} + \dfrac{3}{10} \text{ meter} + \dfrac{1}{4} \text{ meter} + \dfrac{3}{20} \text{ meter} =$ _____

12. $2\dfrac{1}{4} \text{ MW} + 5\dfrac{1}{2} \text{ MW} + 4\dfrac{1}{8} \text{ MW} + \dfrac{9}{16} \text{ MW} =$ _____

13. Technician Makena spent $\frac{1}{2}$ of a workday on a wind turbine repair and $\frac{1}{3}$ of the workday on a new installation. What part of her workday did Makena spend on these two projects? _____

14. Cable is needed to connect the two low-head hydro generators with the distribution panel shown. What is the total amount of cable needed for the project? _____

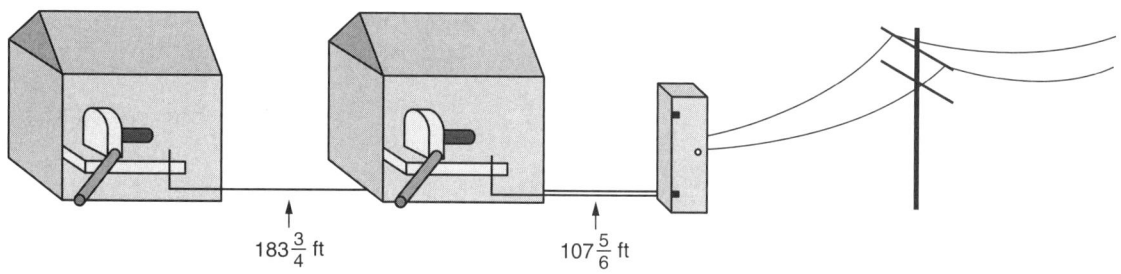

$183\frac{3}{4}$ ft $107\frac{5}{6}$ ft

15. Two PV arrays are installed by a home owner. The array on the house roof provides $\frac{1}{3}$ of the house load. The array on the garage roof provides $\frac{1}{4}$ of the house load. What part of the house load is provided by the PV arrays? _____

16. What is the total length of straight 7-inch diameter pipe needed for the hydro system shown? _____

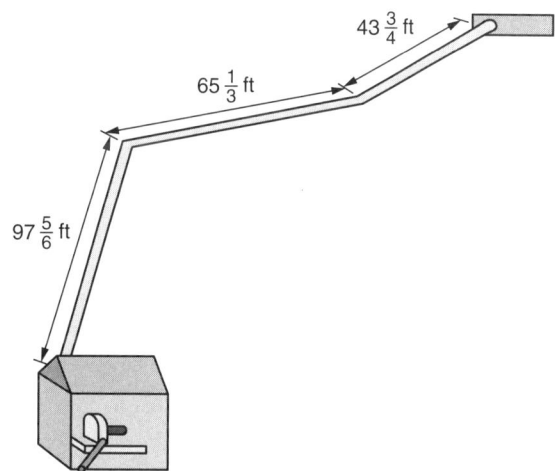

$43\frac{3}{4}$ ft

$65\frac{1}{3}$ ft

$97\frac{5}{6}$ ft

17. Three batches of grout were needed to fill a geothermal well. Those batches were $41\frac{1}{2}$ gallons, $41\frac{5}{8}$ gallons, and $41\frac{1}{4}$ gallons. What was the total volume of grout needed to fill the well? _____

18. Technician Hayden made 3 service calls today. He used $\frac{1}{4}$ of a tank of gas to drive to his first call, $\frac{1}{8}$ of a tank to get to the second call, and $\frac{3}{8}$ of a tank for his third call. What part of the gas in the tank did Hayden use today? _____

19. Four wind turbines need the conduits shown to join the main distribution conduit. How much tubing is needed to make these conduits?   _____

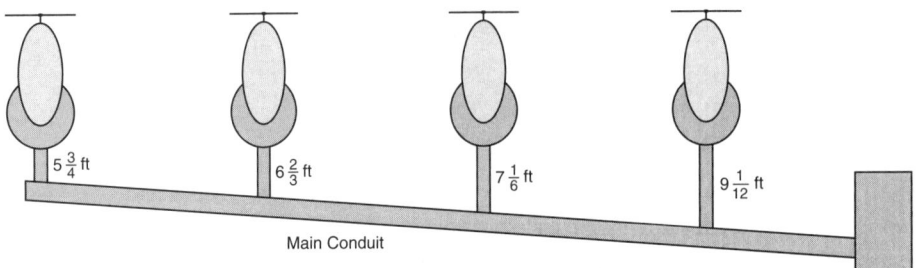

$5\frac{3}{4}$ ft $6\frac{2}{3}$ ft $7\frac{1}{6}$ ft $9\frac{1}{12}$ ft

Main Conduit

20. A wind turbine tower comes in two parts. The parts are $20\frac{1}{3}$ ft and $15\frac{3}{4}$ ft tall. How tall is the tower when the parts are put together?   _____

21. How far is the top of the solar collector above the roof?   _____

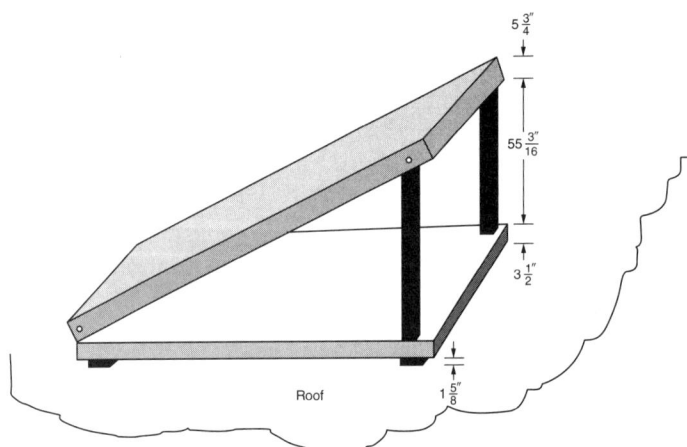

$5\frac{3}{4}''$

$55\frac{3}{16}''$

$3\frac{1}{2}''$

Roof $1\frac{5}{8}''$

22. Three identical solar hot water heater panels are installed on a roof. They are assembled with $\frac{3}{4}$ inch spacers between the panels. What is the total width taken up with this set up?   _____

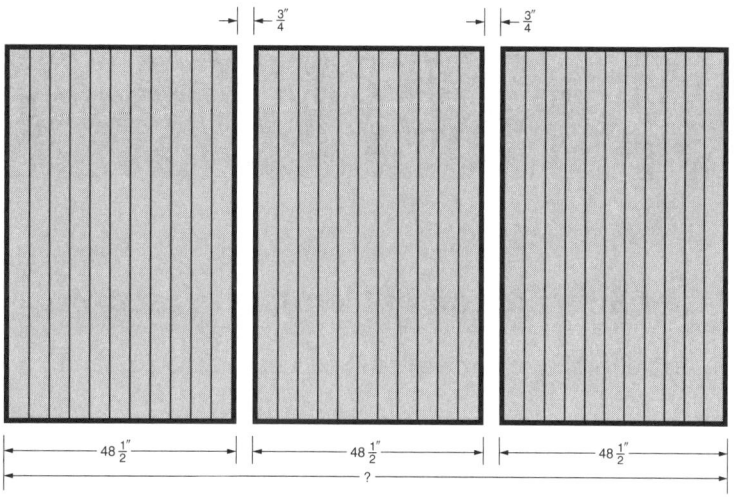

23. The owner of a PV system has analyzed the last 15 years of energy savings to see if the system has paid for itself. The first 5 years totaled a savings of $\frac{1}{3}$ of the cost of the system, the next 5 years produced savings of $\frac{1}{4}$ of the cost, and for the last 5 years, the savings were $\frac{3}{8}$ of its cost. What part of the cost has been saved?   _____

24. The 5-J Construction Company is installing a 3-ton horizontal ground loop heat pump system. Because of the contour of the land, the setup has 3 ground loops covering $\frac{1}{20}$ of an acre, $\frac{1}{18}$ of an acre, and $\frac{2}{45}$ of an acre. What is the total land involved in the ground loops?   _____

25. Technician Parker lives on Go Green Acres. He has a variety of renewable energy systems on his property. He generates $\frac{1}{5}$ of his energy usage from a low-head hydro system, $\frac{1}{3}$ from a wind turbine, $\frac{1}{4}$ from a geothermal system, and $\frac{1}{6}$ from a PV system. What part of Parker's energy usage is provided by renewable energy systems?   _____

UNIT 7

Subtraction of Common Fractions

Basic Principles of Subtraction of Common Fractions

• Study subtraction of denominate numbers in Section I of the Appendix.

Subtracting fractions is very similar to adding fractions. Fractions must have the same denominators and, when they do, subtract the numerators and place the difference over the denominator.

EXAMPLE 1: Subtract $\frac{1}{9}$ from $\frac{5}{9}$.

$$\frac{5}{9} - \frac{1}{9} = \frac{5-1}{9} = \frac{4}{9}$$

The last step is to see if the fraction can be reduced to lower terms. The same process applies when using the vertical form.

EXAMPLE 2: What is the difference between $\frac{7}{11}$ and $\frac{2}{11}$?

$$\begin{array}{r} \frac{7}{11} \\ -\frac{2}{11} \\ \hline \frac{7-2}{11} = \frac{5}{11} \end{array}$$

The answers can also be estimated using a method very similar to estimating while adding fractions.

If the fractions to be subtracted do not have the same denominator, equivalent fractions with a common denominator must be created and then subtracted. This is just like when adding fractions.

EXAMPLE 3: Find the difference between $\frac{5}{8}$ and $\frac{1}{6}$.

The denominators are different, so we need to find a common denominator. 24 is the common denominator. We next make equivalent fractions with 24 as the denominator.

$$\frac{5}{8} - \frac{1}{6} = \frac{5 \times 3}{8 \times 3} - \frac{1 \times 4}{6 \times 4} = \frac{15}{24} - \frac{4}{24} = \frac{15 - 4}{24} = \frac{11}{24}$$

When working with mixed numbers, subtract the fractions first, then subtract the whole numbers. Always do it in that order! The reason is that sometimes the fraction being subtracted is larger than the fraction it is being subtracted from. You have to borrow one from the whole number making it one less, and converting the smaller fraction to an improper fraction. Now the fractions can be subtracted and then the new whole numbers can be subtracted.

EXAMPLE 4: Find the difference between $6\frac{2}{3}$ and $4\frac{5}{6}$.

The common denominator is 6.

$$6\frac{2}{3} - 4\frac{5}{6} = 6\frac{2 \times 2}{3 \times 2} - 4\frac{5}{6} = 6\frac{4}{6} - 4\frac{5}{6}.$$

We cannot subtract 5 from 4. So we borrow 1 from the whole number 6, making it 5. The 1 gets converted to an equivalent fraction with the common denominator

$$6\frac{2}{3} - 4\frac{5}{6} = 6\frac{2 \times 2}{3 \times 2} - 4\frac{5}{6} = 6\frac{4}{6} - 4\frac{5}{6} = 5\frac{4 + 6}{6} - 4\frac{5}{6} = 5\frac{10}{6} - 4\frac{5}{6} = (5 - 4)\frac{10 - 5}{6} = 1\frac{5}{6}$$

This same example could be solved by converting to improper fractions and subtracting, then converting back to mixed numbers. You would obtain the same results.

Remember and use the following guidelines when subtracting common fractions:

* Find the common denominator.
* Make equivalent fractions using the common denominator.
* Reduce the fraction in the answer to the lowest terms.

Practical Problems:

- Apply the principles of subtraction of common fractions to the problems in this unit.

In problems 1–12, subtract the numbers.

1.
$$\frac{7}{9}$$
$$-\frac{2}{9}$$

2.
$$\frac{5}{8}$$
$$-\frac{3}{16}$$

3.
$$5\frac{5}{7}$$
$$-1\frac{1}{7}$$

4.
$$4\frac{3}{5}$$
$$-2\frac{2}{9}$$

5.
$$7\frac{1}{7}$$
$$-\frac{4}{9}$$

6. $\dfrac{8}{11} - \dfrac{5}{11} =$ _____

7. $\dfrac{4}{5} - \dfrac{2}{3} =$ _____

8. $6\dfrac{7}{8} - 4\dfrac{3}{8} =$ _____

9. $3\dfrac{3}{5} - \dfrac{1}{2} =$ _____

10. $4\dfrac{1}{6} - 3\dfrac{2}{3} =$ _____

11. $\dfrac{3}{4}$ feet $- \dfrac{3}{8}$ feet $=$ _____

12. $\dfrac{9}{16}$ pounds $- \dfrac{1}{16}$ pounds $=$ _____

13. $8\dfrac{5}{8}$ inches $- 6\dfrac{3}{8}$ inches $=$ _____

14. $5\frac{1}{3}$ gallons $- 2\frac{1}{2}$ gallons $=$ _____

15. On a service call, technician Brett used $75\frac{1}{4}$ feet of wire from his 200-foot coil on his truck. How much wire is left on the coil? _____

16. What is the size of the gap between the PV collector and the roof shown? _____

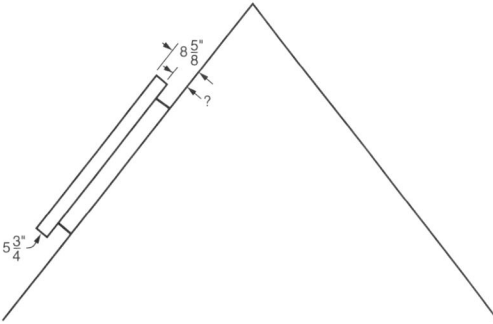

17. A ground source heat pump system has an open loop well that is $575\frac{1}{2}$ feet deep. Tubing collecting water and taking it to the heat pump is inserted into the well to a depth of $557\frac{3}{4}$ feet. How far off the bottom is the opening of the tubing? _____

18. What is the clearance of the blade from the ground for the wind turbine shown? _____

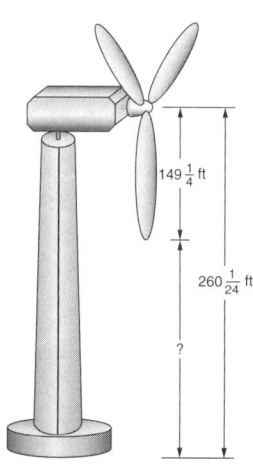

19. A solar hot water collector receives $4\frac{7}{8}$ kW each hour on a specific day. Water removes $4\frac{1}{16}$ kW each hour. The rest is lost. For this day, how much heat is lost each hour and not collected?

20. What is the actual elevation drop for the hydro system shown?   _____

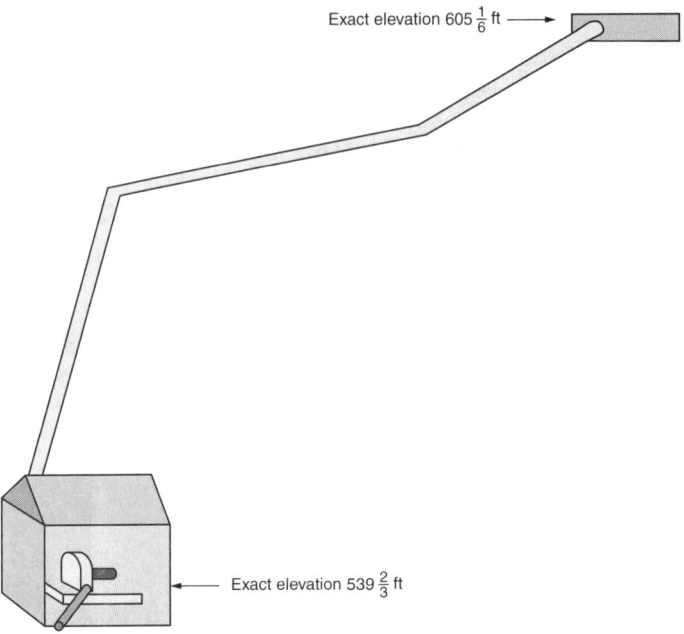

Exact elevation 605 $\frac{1}{6}$ ft →

Exact elevation 539 $\frac{2}{3}$ ft ←

21. What are the dimensions of the active surface of the solar collector shown?

_____

_____

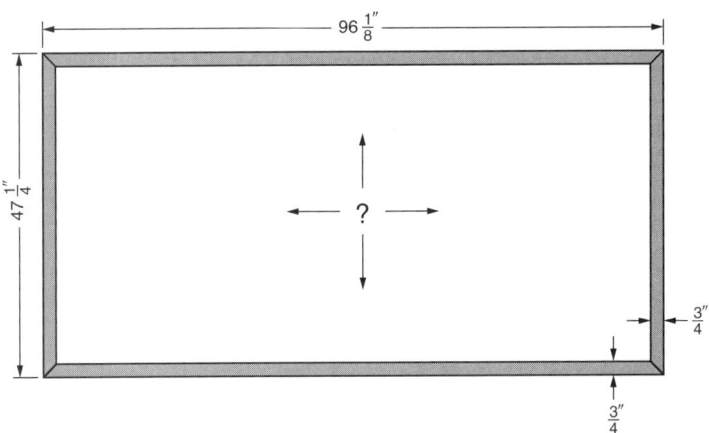

96 $\frac{1''}{8}$

47 $\frac{1''}{4}$

?

$\frac{3''}{4}$

$\frac{3''}{4}$

22. The distance from the center of the hub to the tip of a blade is shown. The distance from the center of the hub to its edge is $3\frac{3}{4}$ ft. Find the length of the blade.

_____

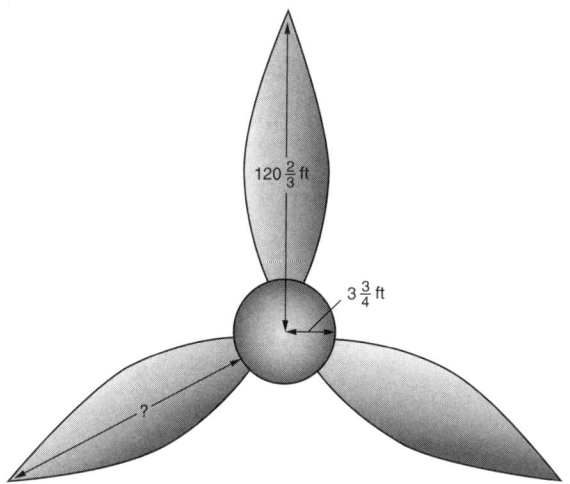

23. A $5\frac{3}{4}$ foot piece of 6-inch diameter pipe is cut from a $10\frac{1}{2}$ foot piece in the stock room. How much 6-inch pipe is left in the stockroom?

24. What is the outside diameter of the pipe used to draw water from a well for an open loop geothermal system?

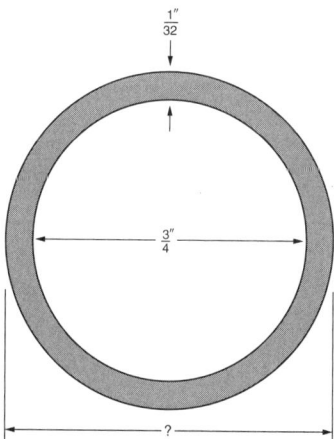

25. On a repair call to fix a bearing on a generator, technician Carson spent $5\frac{7}{12}$ hours. He spent $2\frac{1}{3}$ hours of that time disassembling the low-head hydro generator and $1\frac{1}{2}$ hours reassembling it. How much time did Carson spend making the actual repair to the bearing?

UNIT 8

Multiplication of Common Fractions

Basic Principles of Multiplication of Common Fractions

- Study multiplication of denominate numbers in Section I of the Appendix.

Multiplying fractions is simpler and easier than adding or subtracting fractions. There is no need to find a common denominator. It is easiest to always put the problem in linear form. Then you simply multiply the numerators and multiply the denominators.

EXAMPLE 1: Multiply $\frac{2}{3}$ by $\frac{5}{7}$.

$$\frac{2}{3} \times \frac{5}{7} = \frac{2 \times 5}{3 \times 7} = \frac{10}{21}$$

The last step is to see if the fraction can be reduced to lower terms. When a problem is given in the vertical form, rewrite it in linear form. The reducing to lower terms can be accomplished before multiplying by removing common factors—one in the numerator and the same one in the denominator—before multiplying. This actually makes the multiplication easier. There may be more than one common factor. They all can be removed before multiplying. The removal of the common factors is called cancelling. If you do not cancel before multiplying, you can do it afterward. It is the same thing as reducing to lower terms.

EXAMPLE 2: What is the product of $\frac{7}{8}$ and $\frac{2}{21}$?

$$\frac{\overset{1}{\cancel{7}}}{\underset{4}{\cancel{8}}} \times \frac{\overset{1}{\cancel{2}}}{\underset{3}{\cancel{21}}} = \frac{1 \times 1}{4 \times 3} = \frac{1}{12}$$

The answers can also be estimated using a method very similar to estimating while adding fractions, except you are multiplying rather than adding the estimated numbers.

Another difference between adding and multiplying is that when multiplying mixed numbers, you always want to convert the mixed number to an improper fraction before multiplying. You may then have to convert back to a mixed number (not always). Also, anytime you multiply a whole number by itself times a fraction, you can make an improper fraction of the whole number by putting it over a denominator of 1.

EXAMPLE 3:   Find the product of $12 \times 2\frac{1}{5} \times \frac{1}{18}$.

$$12 \times 2\frac{1}{5} \times \frac{1}{18} = \frac{12}{1} \times \frac{(2 \times 5) + 1}{5} \times \frac{1}{18} = \frac{\overset{2}{\cancel{12}}}{1} \times \frac{11}{5} \times \frac{1}{\underset{3}{\cancel{18}}} = \frac{2 \times 11}{5 \times 3} = \frac{22}{15} = 1\frac{7}{15}$$

Remember that the units associated with the numbers do not have to be the same. Any units associated with the numbers get multiplied together.

Remember and use the following guidelines when multiplying common fractions:

• You do not have to find the common denominator.
• Multiplication problems are best worked in the linear form rather than the vertical form.
• Change all mixed numbers to improper fractions.
• You do not have to cancel, but if you do, one of the numerators and one of the denominators must be divided by the same number.
• Units can be canceled in a similar manner.
• You may cancel more than once.
• Reduce the answer to lowest terms.

Practical Problems:

• Apply the principles of multiplication of common fractions to the problems in this unit.

In problems 1–14, multiply the fractions and mixed numbers.

1. $\frac{1}{3}$

   $\times \frac{2}{3}$

2. $\frac{1}{4}$

   $\times \frac{3}{5}$

3. $2\frac{3}{7}$

   $\times \frac{4}{5}$

4. $3\dfrac{1}{4}$

$\times\ 2\dfrac{1}{5}$

5. $\dfrac{1}{4}$

$\times\ \dfrac{4}{9}$

6. $\dfrac{3}{8}$

$\times\ \dfrac{4}{15}$

7. $4\dfrac{1}{5} \times 1\dfrac{1}{14} =$ _____

8. $1\dfrac{3}{4} \times 1\dfrac{5}{7} =$ _____

9. $\dfrac{1}{3} \times \dfrac{1}{2}$ inch $=$ _____

10. $1\dfrac{1}{8}$ pounds $\times \dfrac{2}{5} =$ _____

11. $5\dfrac{1}{3}$ feet $\times \dfrac{9}{32}$ feet $=$ _____

12. $\dfrac{3}{4} \times \dfrac{3}{25} \times 2\dfrac{2}{9}$ MW $=$ _____

13. $3\dfrac{1}{3}$ feet $\times 12 =$ _____

14. $2\dfrac{1}{4}$ pounds $\times 16\ \dfrac{\text{ounces}}{\text{pound}} =$ _____

15. Find the distance between the first and last wind turbine.   _____

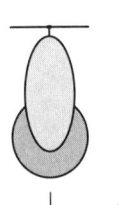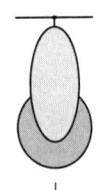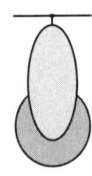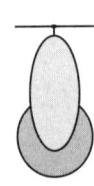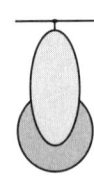

$537\frac{5}{6}$ ft

?

16. What is the length of roof needed to hold the solar collectors? _____

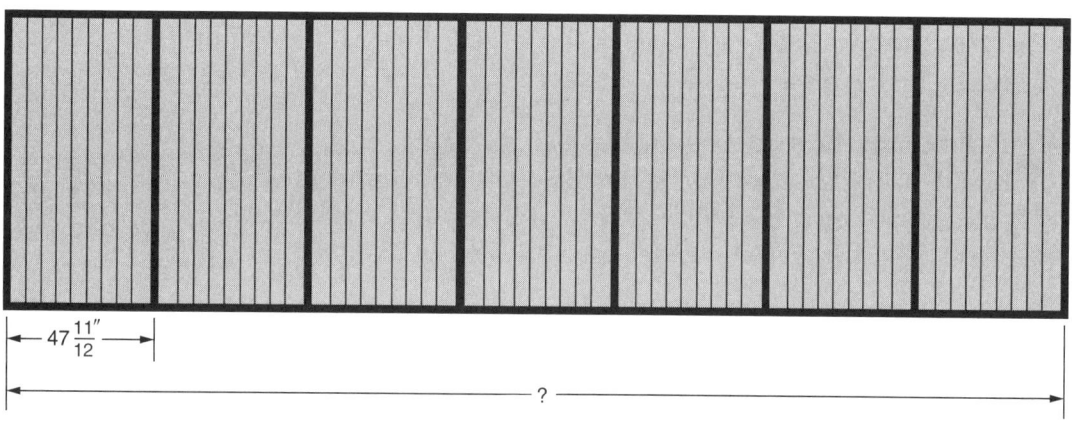

$47\frac{11}{12}''$

?

17. The minimum bending radius for copper tubing is $2\frac{1}{2}$ times the copper tubing diameter. What would the minimum radius be for $\frac{5}{8}$-inch copper tubing? _____

18. In August, the flow in a potential low-head hydro system stream decreases to $\frac{5}{8}$ the flow during April. The flow during April averages $124\frac{1}{6}$ gpm (gallons per minute). What is the average flow during August? _____

19. A small wind turbine, when operating, produces $\frac{3}{5}$ of a house's electrical needs. The house use averages $2\frac{5}{12}$ kW each hour. How much does the wind turbine produce when it operates? _____

20. On a cloudy day, a PV array has an output that is $\frac{5}{8}$ what it would be on a sunny day. If the array produces 400 watts each hour on a sunny day, what is the output on a cloudy day? _____

21. The 5-J Construction Company increased a customer's solar array to $1\frac{3}{4}$ times what it had been. It used to have a collection area of 282 sq ft. What is the new collection area? _____

22. Technician Hayden worked $9\frac{1}{2}$ hours on one service call. He spent $\frac{1}{6}$ of the time diagnosing the problem. How much time did Hayden spend diagnosing the problem? _____

23. Two low-head hydro generators are installed on two streams on a farm. The larger generator produces at a rate of $7\frac{1}{2}$ kW each hour. Because of the differences in flows and elevation drops, the smaller generator produces at a rate $\frac{6}{7}$ times the larger. What is the rate of power generation for the smaller generator?   _____

24. A ground source heat pump system is being installed in a particular location. For that location and soil makeup, it is recommended that $170\frac{1}{2}$ feet of well be drilled for each ton of heat pump capacity. How many feet of well are recommended for a $2\frac{1}{2}$-ton heat pump?   _____

25. The weight of one solar collector panel is $2\frac{3}{4}$ times the weight of the same size collector panel for a PV system. The PV panel weighs $35\frac{1}{4}$ pounds. What is the weight of the solar collector?   _____

UNIT 9

Division of Common Fractions

Basic Principles of Division of Common Fractions

- Study division of denominate numbers in Section I of the Appendix.

Dividing fractions is very similar to multiplying—with one important difference. You first need to invert the divisor (the fraction after the ÷ sign). You invert by making the old denominator the new numerator and the old numerator the new denominator. Then you treat it as a multiplication problem. Be careful to first make improper fractions of all mixed numbers. (All whole numbers have a 1 as their denominator.) Also be very careful to invert first before any cancelling is done.

EXAMPLE 1:   Divide $13\frac{1}{5}$ by $7\frac{1}{3}$.

Estimating the answer: $13 \div 7 = $ a little less than 2.

$$13\frac{1}{5} \div 7\frac{1}{3} = \frac{(13 \times 5) + 1}{5} \div \frac{(7 \times 3) + 1}{3} = \frac{66}{5} \div \frac{22}{3} = \frac{\overset{3}{\cancel{66}}}{5} \times \frac{3}{\underset{1}{\cancel{22}}} = \frac{3 \times 3}{5 \times 1} = \frac{9}{5}$$

The last step is to see if the fraction can be reduced to lower terms.

$$\frac{9}{5} = 1\frac{4}{5}$$

So,

$$13\frac{1}{5} \div 7\frac{1}{3} = 1\frac{4}{5}$$

Remember that the units associated with the numbers do not have to be the same. Any units associated with the numbers get divided just as the numbers did.

Remember and use the following guidelines when dividing common fractions:

- You do not have to find the common denominator.
- Division problems are best worked in the linear form.
- Change all mixed numbers to improper fractions.
- Invert the divisor (including any units) before any cancelling.
- You do not have to cancel, but if you do, one of the numerators and one of the denominators must be divided by the same number.
- You may cancel more than once.
- Reduce the answer to lowest terms.

Practical Problems:

- Apply the principles of division of common fractions to the problems in this unit.

In problems 1–12, divide the fractions and mixed numbers.

1. $\dfrac{5}{11} \div \dfrac{3}{4} =$

2. $\dfrac{7}{9} \div \dfrac{3}{14} =$

3. $\dfrac{5}{32} \div \dfrac{1}{4} =$

4. $\dfrac{15}{16} \div \dfrac{3}{8} =$

5. $3\dfrac{1}{2} \div 3\dfrac{2}{3} =$

6. $2\dfrac{1}{4} \div \dfrac{7}{8} =$

7. $4 \div 1\dfrac{2}{7} =$

8. $13\dfrac{1}{3} \div 2\dfrac{2}{3} =$

9. $\dfrac{3}{8}$ pound $\div\ 6 =$

10. $12\dfrac{2}{5}$ sq ft $\div\ 1\dfrac{6}{25}$ ft $=$

11. $4{,}355\dfrac{2}{3}$ gallons $\div\ 15$ minutes $=$

12. $7\dfrac{7}{8}$ ft $\div\ 2\dfrac{2}{3}$ ft $=$

13. A large-scale wind farm currently has 12 wind turbines out of service. This is $\frac{3}{16}$ of the total on the farm. What is the number of turbines on the farm?

_____

14. A low-head hydro generation system produces 300 watts of power using a flow of $12\frac{1}{4}$ gpm. What is the power produced per gallon per minute of flow?

_____

15. $94\frac{1}{6}$ hours were spent performing identical inspections on 14 identical wind turbines. How much time was spent inspecting each turbine?

_____

16. A cable connecting the output from a PV collector with the main distribution bus is $17\frac{1}{6}$ feet long. How many of these sized cables can be cut from a 200-foot roll of cable?

_____

17. A solar hot water collector is shown. Its length is $119\frac{5}{8}$ in. What is the width of the collector? HINT: Area divided by length gives width.

_____

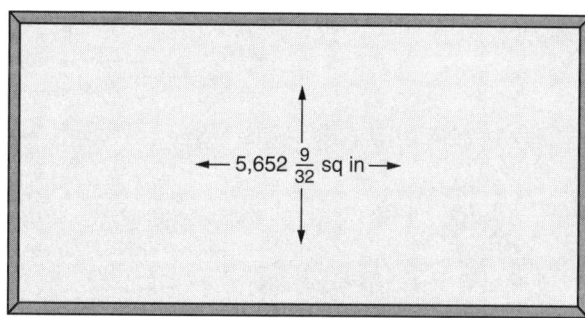

$\longleftarrow 5,652\ \frac{9}{32}$ sq in $\longrightarrow$

18. What is the kW output per hour of a hydro generator if it produces $28\frac{1}{3}$ kWh in running $6\frac{2}{3}$ hrs?

_____

19. A geothermal well drilling rig drills 150 feet in $3\frac{1}{2}$ hours. How far would it drill in 1 hour?

_____

20. The surface area of a hole can be found by dividing the volume of the hole by its depth. The volume of dirt removed from a horizontal exchange field for a geothermal system is $16,740\frac{3}{4}$ cu ft. The depth of the field is $6\frac{1}{2}$ ft. What is the surface area of the field?

_____

21. A PV collector produces $14\frac{2}{3}$ watts per sq ft of collector. How big would the collector have to be to produce $266\frac{2}{3}$ watts?

_____

22. A geothermal well requires $\frac{2}{3}$ gallon of grout for each foot of well. One well required $352\frac{3}{4}$ gallons of grout. How deep is the well?

_____

23. A low-head hydro generator produces $2{,}955\frac{3}{4}$ watts as a result of $34\frac{1}{3}$ gallons of water flowing through the system. What is the rate of power production for the generator in watts per gallon?

_____

24. A solar collector has a tall building cast a shadow on it so that it only receives sunlight $\frac{3}{4}$ of the time. If it collected $23{,}785\frac{5}{7}$ Btus one day, what would it have collected if the building were not there?

_____

25. Technician Parker spent 7 hours installing a penstock for a low-head hydro generator. He determined that this was $\frac{2}{5}$ of the total time installing the generator system. What was Parker's total time installing the system?

_____

UNIT 10

Combined Operations with Common Fractions

Basic Principles of Combined Operations With Common Fractions

- Review and apply the principles of addition, subtraction, multiplication, and division of common fractions to the problems in this unit.
- Read each problem. Decide which operation must be performed to solve the problem. (The same hints given in Unit 5 can be used with these problems to help determine which operation to perform.)
- Perform the operation. Some problems may require the application of more than one type of operation to solve the problem. As a result, it may take more than one step to find the answer.

EXAMPLE 1: Replacing the top surface of an installed solar hot water collector takes 2 technicians a total of 16 man-hours. Technician Hayden spent $\frac{2}{3}$ of that time working on the job. How much time did the other technician, Parker, spend on the job?

This problem is solved using multiple operations. To estimate the answer, we should realize that $\frac{2}{3}$ is larger than $\frac{1}{2}$.

$$\frac{1}{2} \times 16 \text{ man-hours} = \frac{1 \times \overset{8}{\cancel{16}}}{\underset{1}{\cancel{2}} \times 1} \text{ man-hours} = \frac{8}{1} \text{ man-hours} = 8 \text{ man-hours.}$$

Hayden should have worked more than 8 man-hours. Parker should have worked less than $16 - 8 = 8$ man-hours.

To solve the exact problem, first, multiply $\frac{2}{3} \times 16$ man-hours $= \frac{2 \times 16}{3}$ man-hours $= \frac{32}{3}$ man-hours $= 10\frac{2}{3}$ man-hours. This is the time that Hayden worked on the job. For Parker's time, 16 man-hours $- 10\frac{2}{3}$ man-hours $= 15\frac{3}{3}$ man-hours $- 10\frac{2}{3}$ man-hours $= (15 - 10)\frac{3-2}{3}$ man-hours $= 5\frac{1}{3}$ man-hours.

Hayden's time is greater than 8 hours, and Parker's time is less than 8 hours, just as was estimated.

Practical Problems:

1. $\begin{array}{r} \frac{3}{4} \\ + \frac{5}{9} \\ \hline \end{array}$

2. $\begin{array}{r} 4\frac{1}{7} \\ 2\frac{4}{9} \\ + 3\frac{17}{21} \\ \hline \end{array}$

3. $\begin{array}{r} \frac{9}{16} \\ - \frac{1}{4} \\ \hline \end{array}$

4. $\begin{array}{r} 3\frac{1}{9} \\ - 1\frac{5}{18} \\ \hline \end{array}$

5. $\begin{array}{r} \frac{4}{5} \\ \times \frac{1}{2} \\ \hline \end{array}$

6. $\begin{array}{r} 5\frac{1}{5} \\ \times 1\frac{7}{13} \\ \hline \end{array}$

7. $\frac{4}{7} \div \frac{8}{21} =$

8. $3\frac{2}{3} \div 3\frac{1}{7} =$

9. $1\frac{2}{5}$ gallons $+ 3\frac{6}{7}$ gallons $=$ _____

10. $5\frac{2}{3}$ hours $- 2\frac{5}{6}$ hours $=$ _____

11. $4\frac{1}{2}$ inches $\times$ $2\frac{1}{3}$ inches = _____

12. 7 pounds $\div$ $\frac{3}{14}$ sq in = _____

13. A solar hot water collector converts $\frac{7}{10}$ of the sun's energy to heat for the water. If the sun is directing $5\frac{1}{3}$ kW each hour of this day, how much energy does the water gain each hour? _____

14. The warehouse at Nell's wind farm has three partially filled spools of #4/0 wire. The lengths still on the spools are $76\frac{2}{3}$ ft, $55\frac{1}{12}$ ft, and $39\frac{5}{6}$ ft. If spliced together, what length wire would be made? _____

15. One home's geothermal system needs $437\frac{1}{2}$ ft of well depth. $258\frac{2}{3}$ ft have already been drilled. What is the well depth that still needs to be drilled? _____

16. A wind turbine had a ground-to-blade tip clearance of $20\frac{1}{3}$ ft. After a rainy season, the turbine tower settled $\frac{1}{4}$ ft into the ground. What is the new ground-to-blade tip clearance? _____

17. How far does the ladder extend out past the back end of the repair van? _____

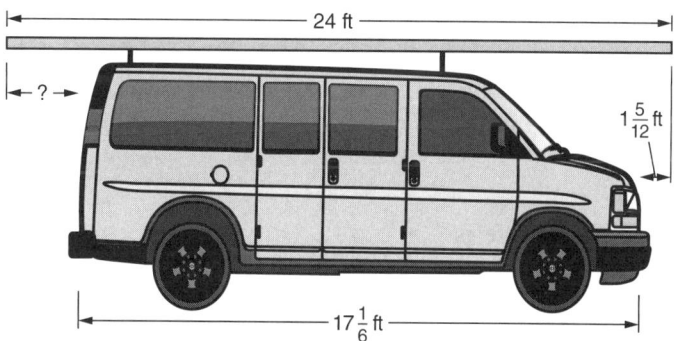

18. How much electrical energy is created by running a $5\frac{1}{2}$ kW low-head hydro generator for $3\frac{1}{12}$ hours? _____

19. PV disks occupy $1\frac{3}{16}$ of the area of a collector frame. If the frame is $39\frac{1}{4}$ sq ft in area, what is the actual area of the PV disks? _____

20. How far off the ground is the penstock at the point indicated in the diagram? _____

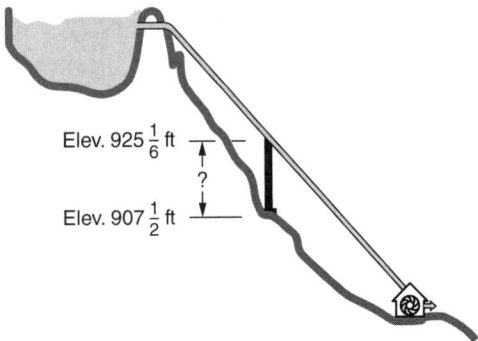

Elev. 925 $\frac{1}{6}$ ft

?

Elev. 907 $\frac{1}{2}$ ft

21. Twelve identical wind turbines generated 23,101$\frac{1}{5}$ kWh for one day. What was the generation per turbine? _____

22. A hole drilled in the ground creates a volume measured by surface area times depth. Well depth equals hole volume divided by surface area. One well has a volume of 70$\frac{1}{2}$ cu ft and a surface area of $\frac{1}{5}$ sq ft. What is the depth of the well? _____

23. The time for a quantity of water to flow past a point can be found by dividing the quantity of water by the flow rate. A stream has a flow rate of 92$\frac{5}{6}$ gps. How long will it take for 2,367$\frac{3}{4}$ gallons to flow past the point? _____

24. The area of a roof on a house is 236$\frac{1}{2}$ sq ft. Only $\frac{1}{8}$ of the roof is *not* covered with solar collectors. What is the area of the solar collectors on the roof? _____

25. The circulation pump of a solar collector system ran 5$\frac{2}{3}$ hours, 6$\frac{1}{2}$ hours, 6$\frac{1}{12}$ hours, 2$\frac{1}{4}$ hours, 5$\frac{3}{4}$ hours, 6$\frac{3}{4}$ hours, and 5$\frac{5}{12}$ hours one week. Suppose the pump had to run the same amount of time on each day. How long would the pump have to run on each day to get the same total as it has this week? _____

Decimal Fractions

UNIT 11

Addition of Decimal Fractions

Basic Principles of Addition of Decimal Fractions

- Study addition of denominate numbers in Section I of the Appendix.

A decimal number is composed of two parts separated by a decimal point. The part to the left of the decimal point is the whole number part and the part to the right is the fraction part. The whole numbers studied in Section 1 had no fraction part to them. When there is no fraction part, the decimal point is left off of the number. Although they are properly called decimal fractions, most people refer to these numbers as decimals.

Just as each number position in whole numbers has a name, each position in a decimal fraction has a name. The names for the fraction positions are similar to and follow the whole number names with one exception. There is no name related to the units name. The number 123.4567809 has the whole number part as 123 with 1 in the hundreds position, 2 in the tens position and 3 in the units position. The decimal fraction part of the number has 4 in the ten<u>ths</u> position, 5 in the hundred<u>ths</u> position, 6 in the thousand<u>ths</u> position, 7 in the ten-thousand<u>ths</u> position, 8 in the hundred-thousand<u>ths</u> position, 0 in the million<u>ths</u> position, and 9 in the ten-million<u>ths</u> position. The *th* is underlined here just to emphasize that adding it to a position name indicates that the number is a decimal fraction position. Notice that the order is the same as for whole number positions moving away from the decimal point (ignoring the units position): ten, hundred, thousand, ten thousand, hundred thousand, million, and so on.

One difference with decimal fractions compared with whole numbers is that no commas are used with the fraction part. If the number is only a decimal fraction, a 0 is always written in the units position.

When adding whole numbers, care must be taken to add units to units and tens to tens and so on. This is done by lining up the units column when putting the numbers in vertical alignment. A similar arrangement must be done when adding decimal fractions, but this is easily accomplished by lining up the decimal points. If this is done, the columns are aligned. If adding a whole number and a decimal fraction, the decimal point is after the units number on the whole number. You estimate the answer and add as we had done with whole numbers.

EXAMPLE 1:   Add 113.523, 57.6, and 662.19.

First, estimate the answer.

We are estimating the answer by adding 100, 60, and 700.

$$
\begin{array}{r}
100 \\
60 \\
+\ 700 \\
\hline
860
\end{array}
$$

Now, add the actual numbers. Line up the decimal points.

$$
\begin{array}{r}
\overset{111\ 1}{113.523} \\
57.6 \\
+\ 662.19 \\
\hline
833.313
\end{array}
$$

Just as we do not put zeros to the left of whole numbers, we do not put zeros to the right of decimal numbers.

EXAMPLE 2:   What is the sum of 0.4 pounds and 0.87 pounds?

Estimating, we add 0.4 pounds and 0.9 pounds.

$$
\begin{array}{r}
0.4 \text{ pounds} \\
+\ 0.9 \text{ pounds} \\
\hline
1.3 \text{ pounds}
\end{array}
$$

Notice that even though neither number had any quantity to the left of the decimal point, the tenth number gets carried over and becomes a whole number.

Finding the actual sum.

$$0.4 \text{ pounds}$$
$$+\ 0.87 \text{ pounds}$$
$$1.27 \text{ pounds}$$

Remember and use the following guidelines when adding decimal fractions:

• Align the decimal points.

Practical Problems:

• Apply the principles of addition of decimal fractions to the problems in this unit.

In problems 1–12, add the numbers.

1.   23.573
   + 56.312

2.   7.843
      5.392
   + 9.055

3.   31.000
      2.485
   + 118.030

4.   7.215 sq ft
      0.937 sq ft
      21.044 sq ft
   + 34.886 sq ft

5.   123.44 MW
      84.715 MW
   +   4.345 MW

6.   67.908 gallons
      851.7   gallons
   + 348.263 gallons

7. 17.362 + 34.835 =   _____

8. 1.423 + 723.25 + 81.818 =   _____

9. 6.002 + 433.51 + 7.2 =   _____

10. 31.6 hours + 27 hours + 0.55 hours + 9.4 hours =   _____

11. 5.413 cu m + 17.62 cu m + 12.845 cu m + 8.4661 cu m =   _____

12. 83.285 gpm + 121.473 gpm + 44.6953 gpm =   _____

13. During one month, technician Carson worked on wind turbine serial number #435280 for the following times and reasons: 4.5 hours to replace brushes, 27.25 hours for annual inspection and general maintenance, 13 hours to replace a worn bearing, and 6.6 hours to locate and tighten a loose connection. How many hours was Carson involved with turbine #435280 for that month?

_____

14. A batch of grout is created by mixing 50 lb of grout with 295.5 lb of sand and 158.8 lb of water. What is the weight of that batch?

_____

15. Parts needed by technician Hayden when installing a low-head hydro generation system cost as follows: generator - $11,915.92, 135.7 feet of 6-inch straight pipe - $370.62, 4 right angle 6-inch elbows - $55.56. What was the cost of the parts Hayden needed?

_____

16. On one day, the technicians of Suzanne's Energy Service drove their repair trucks 21.7 miles, 19.8 miles, 15.2 miles, 24.5 miles, and 31.9 miles. What was the total mileage driven by Suzanne's technicians that day?

_____

17. A geothermal horizontal ground loop system is shown. How long is the return line to the heat exchanger?

_____

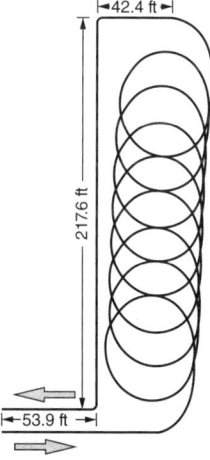

18. A PV system generated the following energy quantities for one sunny week: 15,372.5 kWh, 17,028.3 kWh, 10,879.8 kWh, 2,838.7 kWh, 15,845.5 kWh, 12,771.9 kWh, and 14,899.1 kWh. What was the total energy generation for that week?

_____

19. The FAA (Federal Aviation Administration) requires blinking lights on towers that exceed 200 feet above the ground. Does the wind turbine shown require a blinking light?   _____

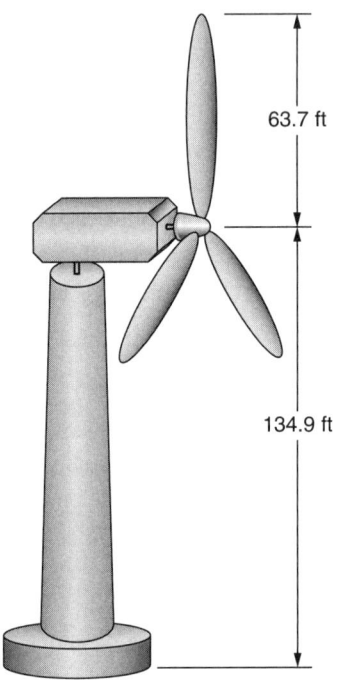

63.7 ft

134.9 ft

20. To install a wind turbine, technician Brett used the following lengths of #4/0 wire: 245.75 m, 193.66 m, and 5.27 m. What is the total length of #4/0 wire used by Brett?   _____

21. Three solar hot water systems were installed. The technicians took 43.5 man-hours, 27.25 man-hours, and 56.33 man-hours to complete the installations. Find the total man-hours spent on the installations.   _____

22. A stream is monitored as a possible source for low-head hydro generation. Five hourly flow samples measured flow rates of 1,093.52 gpm, 1,111.87 gpm, 997.16 gpm, 1,009.94 gpm, and 1,107.33 gpm. What is the total of these flow rates?   _____

23. Two deep wells of 352.5 ft and 277.75 ft were drilled for a geothermal system. What is the total well depth drilled? _____

24. Wind farm owner Nell had a generation capacity of 17.35 MW. Turbines totaling 4.55 MW are added to the farm. What is the new capacity of the farm? _____

25. What is the total length of penstock needed for the low-head hydro system shown? _____

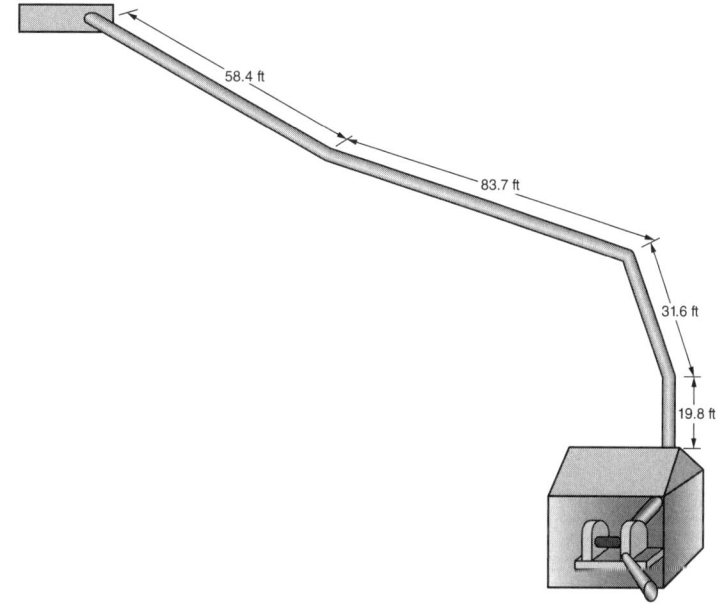

58.4 ft

83.7 ft

31.6 ft

19.8 ft

UNIT 12

Subtraction of Decimal Fractions

Basic Principles of Subtraction of Decimal Fractions

- Study subtraction of denominate numbers in Section I of the Appendix.

To subtract decimals, the decimal points are aligned and a process similar to subtracting whole numbers is followed. One difference is seen when a number is subtracted from a blank; a 0 is placed in the blank, a 1 can be borrowed, making the 0 a 10. Subtraction can then be performed.

EXAMPLE 1:   Subtract 7.44 from 9.8.

First, estimate the answer.

We are estimating the answer by subtracting 7 from 10.

$$
\begin{array}{r}
10 \\
- \ 7 \\
\hline
3
\end{array}
$$

Now, subtract the actual numbers. Line up the decimal points.

$$
\begin{array}{r}
\overset{7\,10}{9.\cancel{8}} \\
- \ 7.44 \\
\hline
2.36
\end{array}
$$

Remember and use the following guidelines when subtracting decimal fractions:

- Align the decimal points.

66

Practical Problems:

• Apply the principles of subtraction of decimal fractions to the problems in this unit.

In problems 1–12, subtract the numbers.

1. $\begin{array}{r} 75.42 \\ -\ 32.11 \\ \hline \end{array}$

2. $\begin{array}{r} 143.271 \\ -\ 117.823 \\ \hline \end{array}$

3. $\begin{array}{r} 572.69 \\ -\ 288.613 \\ \hline \end{array}$

4. $\begin{array}{r} 1,348.036 \text{ sq ft} \\ -\quad 827.91 \ \text{ sq ft} \\ \hline \end{array}$

5. $\begin{array}{r} 5,427 \quad \text{pounds} \\ -\ 2,382.41 \text{ pounds} \\ \hline \end{array}$

6. $\begin{array}{r} 463.881 \text{ gallons} \\ -\ 177.943 \text{ gallons} \\ \hline \end{array}$

7. $56.89 - 24.63 =$ _____

8. $643.299 - 175.38 =$ _____

9. $500 - 169.471 =$ _____

10. $43.74 \text{ MW} - 16.88 \text{ MW} =$ _____

11. $287.3 \text{ hours} - 19.275 \text{ hours} =$ _____

12. $88.47 \text{ meters} - 84.29 \text{ meters} =$ _____

13. A homeowner's house used 523.2 kWh one month. A wind turbine on the property generated 347.65 kWh during that month. How much energy does the homeowner have to buy from the power company? _____

14. A PV system installation is projected to take 72 man-hours. Technician Brett has already worked 17.25 man-hours on the job. How many man-hours must be worked to complete the installation? _____

15. Technician Hayden put in 11.75 man-hours on one repair job. He spent 7.85 man-hours doing the actual repair, the rest of the time he spent getting the equipment ready and restoring it. How much time did Hayden spend getting the equipment ready and restoring it? _____

16. For proper operation of a low-head hydro generation system, the water must have an elevation difference of 8.5 meters. How high must the water in the dam be to meet this minimum requirement?   _____

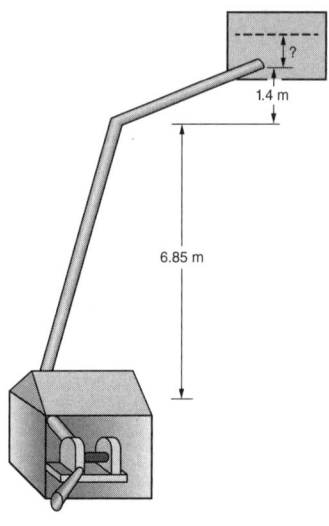

17. The installation of a small wind turbine costs $14,875.90. If the owner prepares and pours the concrete pad for the tower, the cost is $12,880.25. How much would the owner save by pouring his own pad?   _____

18. A property owner has 2 wind turbines on the property. Last month the turbines produced a total of 708.26 kWh of energy. The larger turbine produced 380.82 kWh. How much energy was produced by the smaller turbine?   _____

19. Technician Makena had a 127.75-meter coil of wire on her truck. She cut a 48.9-meter piece from the coil for a repair job. How much wire is left on the coil?   _____

20. Suzanne's warehouse has 875.25 pounds of grout mix. For one geothermal system, 136.6 pounds were used. How much grout mix is left in the warehouse?   _____

21. What area of the house roof will not be covered by the solar collector? _____

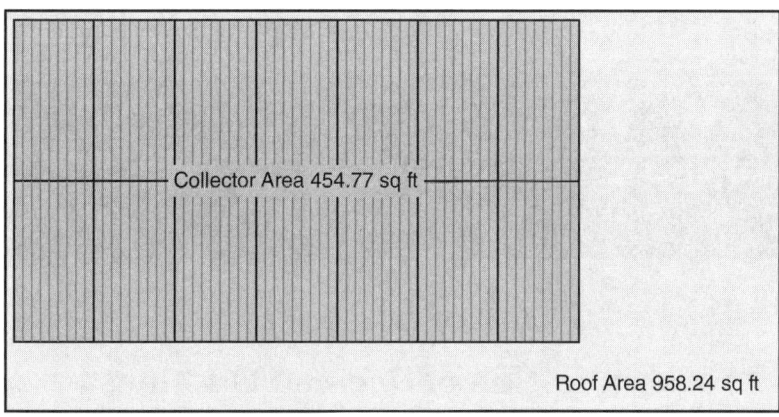

Collector Area 454.77 sq ft

Roof Area 958.24 sq ft

22. A low-head hydro generator is recommended to be serviced every 1,000 hours. It has operated 791.4 hours. How many hours can it operate before it needs to be serviced? _____

23. A 2-ton heating/cooling unit needs 1,100 feet of well for the geothermal system. One well is 653.7 feet deep. How deep must the second well be? _____

24. A repair bill for parts and labor is $311.07. The parts cost $118.98. How much did the labor cost? _____

25. Find the height of the second section of the tower. _____

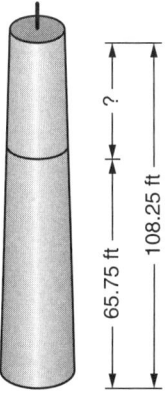

65.75 ft

108.25 ft

?

UNIT 13

Multiplication of Decimal Fractions

Basic Principles of Multiplication of Decimal Fractions

- Study multiplication of denominate numbers in Section I of the Appendix.

When multiplying decimal numbers, line up the last numbers without worrying about where the decimal point is located. Multiply the numbers and get the numerical answer.

EXAMPLE 1: Find the product of 7.13 and 0.6.

Estimating the answer we have 7 times $1 = 7$.

Doing the actual problem we write

$$
\begin{array}{r}
7.13 \\
\times\,0.6 \\
\hline
4278
\end{array}
$$

Now count the number of decimal places in the two numbers. $2 + 1 = 3$. Count three places from the right side of the answer. This places the decimal point between the 4 and the 2, giving an answer of 4.278. (This is close to our estimate of 7.)

If the answer has a 0 as the very last number, it must be counted when placing the decimal point in the answer, but does not have to be written when writing the answer.

Remember and use the following guidelines when multiplying decimal fractions:

- Align the last digits regardless of where the decimal point is located.
- Multiply the numbers.

- Count the total number of decimal places in the problem.
- Count that number from the right end of the answer and place the decimal point.
- Units must be multiplied just as they were when multiplying fractions.

Practical Problems:

- Apply the principles of multiplication of decimal fractions to the problems in this unit.

In problems 1–12, multiply the numbers.

1. 64.3	2. 930.66	3. 0.0782
× 4.77	× 586.4	× 6.433
4. 429.4	5. 257.38 pounds	6. 8.37 MW
× 183.5	× 0.0059	× 41.5
7. 38.65 ft	8. 579.07 gpm	
× 17.5 ft	× 4.062 min	

9. 914.23 × 2.25 = _____

10. 0.084 × 0.0064 = _____

11. $49.27 × 3 = _____

12. 2.018 m × 8.42 = _____

13. A wind farm averaged 113,534.4 kWh for each turbine during 1 month. What was the energy output of the farm that month, if the farm had 27 turbines on it? _____

14. Area is length times width. What is the surface area for the solar collector shown? _____

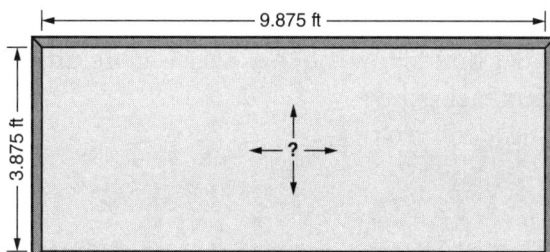

15. Geothermal well grout weighs 15.15 pounds for each gallon. What would be the weight of the grout for one well if it needed 483.75 gallons of grout?   _____

16. Volume equals length times width times depth. How many cubic feet of dirt must be removed to create this horizontal geothermal heat sink?   _____

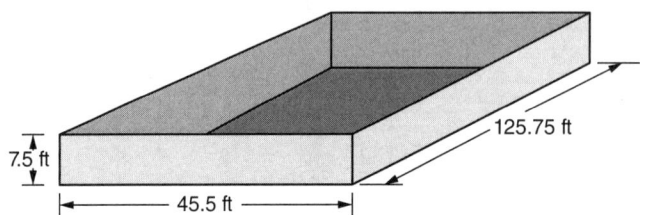

17. An electric company charges $0.06925 per kWh. One month a house used 3,192 kWh. What was the electric charge for that month? (Round off to the next higher cent.)   _____

18. One PV collector costs $513.85. How much would an array of 15 collectors cost?   _____

19. Technician Carson is paid $35.75 an hour. How much does he earn for a 6.25-hour repair job?   _____

20. Technician Makena wants to replace a low-head hydro generator with a larger one. The new one needs flow that is 1.45 times the original flow. If the original flow going through the generator was 2,125 gpm, what flow is needed for the new generator?   _____

21. Each square foot of a solar collector exchanges 35.36 Btu of heat energy with each gallon of water passing through the exchanger. How much energy is given to 4.2 gallons passing through the collector each hour?   _____

22. A low-head hydro generator produces 0.0526 W for each gallon of water flow through it per hour. How much power is generated with a flow of 152,075.5 gallons each hour?   _____

23. When drilling, a rig can drill at a rate of 38.5 feet per hour. How deep can a well be drilled in 5.25 hours?

_____

24. The solar disks of a PV collector cover 0.83 of the total collector area. What is the area of the disks if the collector area is 23.65 sq ft?

_____

25. A stream has a flow rate of 2,536.367 gpm. How many gallons would flow past a point in one day?

_____

UNIT 14

Division of Decimal Fractions

Basic Principles of Division of Decimal Fractions

• Study division of denominate numbers in Section I of the Appendix.

Dividing decimal numbers is very similar to dividing whole numbers. The difference is what to do with the decimal point. Once the decimal point location has been determined, the problem is worked just as the division of whole numbers is worked.

Remember that the whole number division problem is written as

$$\text{DIVISOR } \overline{)\text{DIVIDEND}}^{\text{QUOTIENT}}$$

When the divisor has a decimal, move it to the right to make the divisor a whole number. Move the decimal point in the dividend the <u>same</u> number of places to the right and then place a decimal point in the quotient directly above that place.

EXAMPLE 1:   Divide 65.025 by 4.25.

Estimating the answer we have 60 divided by 4. This gives 15 for our estimated answer.

Dividing the actual numbers gives

$$4.25\,\overline{)65.025}$$

74

Move the decimal point in the divisor two places to the right to make the divisor a whole number. We must move the decimal point in the dividend two places to the right also and then place it in the quotient.

$$4.25\overline{)65.025}$$

Now do the division

```
            15.3
4.25 )65.02 5
      42 5
      22 52
      21 25
       1 27 5
       1 27 5
```

If there are not enough decimal places to move the decimal point in the dividend, add 0s to the end of the dividend.

EXAMPLE 2: Divide 73.75 by 0.0118.

The estimated answer is $70 \div 0.01 = 7{,}000 \div 1 = 7{,}000$. We moved the decimal point two places to the right to make the divisor a whole number, so we move the decimal place two places to the right in the dividend. Two 0s had to be added to 70 to move the decimal point.

Working with the actual problem

$$0.0118\overline{)73.75}$$

The decimal point must be moved four places to the right in the divisor and then also in the dividend. Since there are not four decimal places in the dividend, we need to add two 0s.

$$0.0118\overline{)73.7500}$$

Dividing we get

$$
\begin{array}{r}
6250. \\
0.0118\,)\overline{73.7500} \\
\underline{70\ 8} \\
2\ 95 \\
\underline{2\ 36} \\
590 \\
\underline{590} \\
0 \\
\underline{0}
\end{array}
$$

There are times when the division will not come out evenly or when the answer does not have to be more than a certain number of decimal places. The answer should be rounded off in those cases.

To round off, carry the division one place further than is asked for in the answer (hundredths if tenths were called for, thousandths if hundredths were called for, and so on). Look at the last number. If it is less than 5, drop the last number and you have your answer. If the number is 5 or larger, drop the last number but raise the previous number by one to give you your answer. For example, 3.24 rounded off to the nearer tenth would be 3.2 (because 4 is less than 5); 151.368 rounded off to the nearer hundredth would be 151.37 (8 is larger than 5, so drop the 8, but make the 6 a 7).

Remember and use the following guidelines when dividing decimal fractions:

- Move the decimal point in the divisor.
- Move the decimal point the same number of places to the right in the dividend and place it in the quotient above that point.
- Divide and round off as necessary.

Practical Problems:

- Apply the principles of division of decimal fractions to the problems in this unit.

In problems 1–12, divide the numbers.

1. $37.4 \overline{)987.36}$

2. $0.053 \overline{)38.45998}$

3. $84.29 \overline{)4.357793}$

4. Round off to 3 significant figures:

   $0.0072 \overline{)0.0000048}$

5. Round to the nearer thousandth

   $6,264.32 \overline{)4,643,740.4 \text{ MW}}$

6. $0.00076 \overline{)0.0030172} \text{ pounds}$

7. $4.29 \overline{)0.0290347 \text{ ft}}$

8. Round off to 3 significant figures:

   $0.155 \overline{)0.096565 \text{ m}}$

9. $0.027531 \div 0.057 =$ _____

10. $1,536.112 \div 619.4 =$ _____

11. Round off to 3 significant figures: $0.79064 \text{ miles} \div 0.387 =$ _____

12. Round off to 4 significant figures: $684.42066 \text{ sq ft} \div 7.91 =$ _____

13. A low-head hydro generator produced 3.6 kWh in 12.5 hours. What did the generator produce in one hour? _____

14. How many 12.5-foot pieces of #8 cable can be cut from a 300-foot coil? _____

15. Technician Parker was paid $850.20 for working 27.25 hours on a repair job. What is Parker's hourly rate for this job? _____

16. A solar collector has an area of 37.25 sq ft. In one hour it collected 5,392.352 Btus of heat. How much heat is collected per sq ft of collector? (Round off to 2 decimal places.) _____

17. A penstock for a low-head hydro system is shown. How many feet does the upper section of the penstock lower for each foot of length? (This is the slope of that portion of the penstock.)

_____

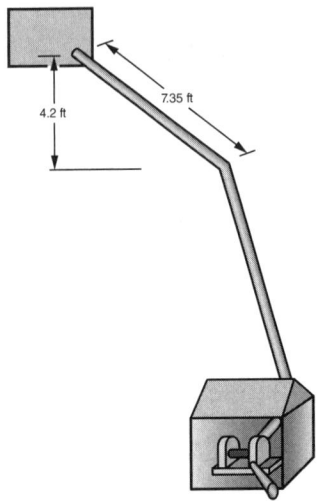

7.35 ft

4.2 ft

18. Conduit used for cables on Nell's Wind Farm cost $544.79. The generation system used 173.5 ft of conduit. What was the cost per foot of conduit on Nell's wind farm?

_____

19. So there is no interference in the wind flow, a minimum area based on the size of the turbine blade around a wind turbine cannot have another turbine built on it. For a particular turbine, the restricted area around the turbine is shown in the diagram. How many turbines might be sited on a plot of land 2 acres in size?

_____

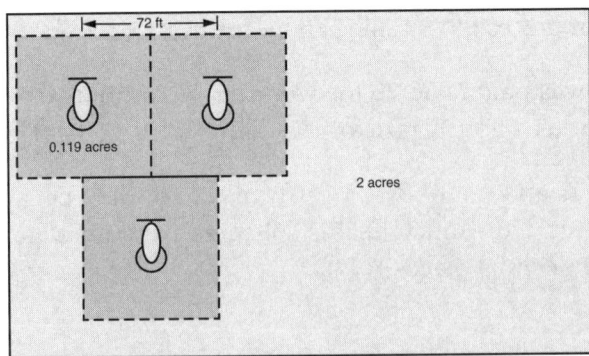

72 ft

0.119 acres

2 acres

20. Technician Hayden serviced 12 identical wind turbines for annual maintenance. He spent the same amount of time on each turbine. If Hayden spent a total of 45 hours on the service call, how much time was spent on each turbine?

_____

21. There are 18 solar collectors on a PV panel. The panel generated 4.1895 kW of power. How much power did each panel contribute to the total?

_____

22. A solar collector panel has the area and length shown. If the width of a rectangle can be found by dividing the area by the length, what is the width of this panel?

_____

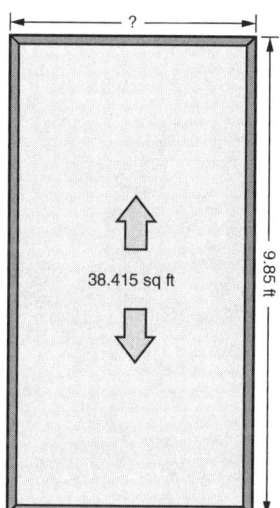

38.415 sq ft

9.85 ft

23. To make grout for a geothermal well, 21.5 gallons of water are used to make 597.5 pounds of grout. How many gallons of water are needed for one pound of grout? (Round off to 4 significant figures.)

_____

24. One low-head hydro generator produces 0.3 kW using a water flow of 633.6 gpm. How much power does it produce from the flow of one gallon per minute? (Round off to 2 significant figures.)

_____

25. The power cable from a wind turbine on top of a ridge needs a pole for support every 53.75 m. How many poles are needed for the arrangement shown? Round to the next higher whole number.   _____

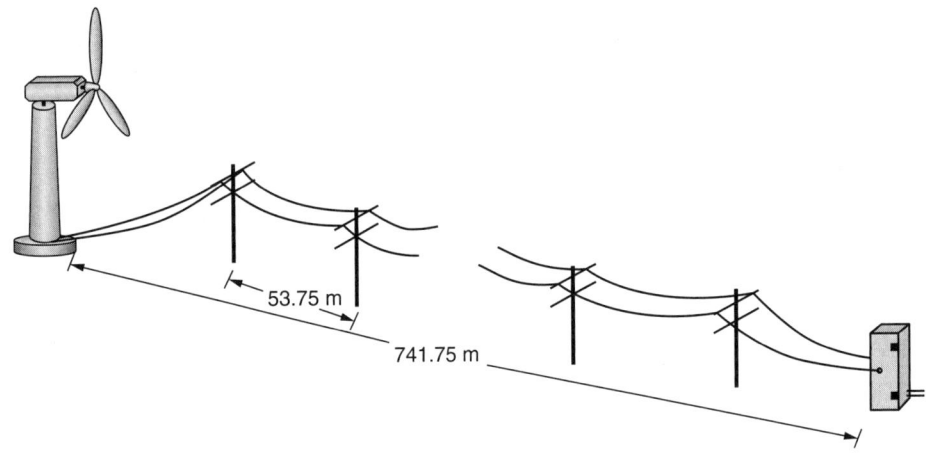

UNIT 15

Decimal and Common Fraction Equivalents

Basic Principles of Decimal and Common Fraction Equivalents

• Study denominate numbers in Section I of the Appendix.

Decimal fractions and common fractions are both types of fractions. Both are parts of a whole number. There should be some way to get from one type to the other.

Problems involving common fractions are best worked out longhand. The main reason for this is that many common fractions do not have exact decimal fraction values. So when some common fractions are converted to decimal fractions on a calculator, the values a calculator uses are not exact, so the final answer is not exactly correct. In other words, you will get a slightly different answer by using a calculator than by working the problem by hand. A second problem with using calculators is that often the answer must be expressed as a common fraction. So in order to work common fraction problems using a calculator, you must first convert the fraction to a decimal, work the problem, and then convert the answer back to a common fraction.

Common fractions related to measurements using a ruler (8ths, 16ths, 32nds, and 64ths) all have exact decimal equivalents. Problems involving these numbers can be worked exactly using a calculator. To get the decimal fraction back to a common fraction, it is easiest to look up the equivalent in a table. The table in Section II of the Appendix contains the equivalents for fractions involving halves, quarters, eighths, and sixteenths.

To convert from a decimal to a common fraction, take the number to the right of the decimal point and make that the numerator of the common fraction. The denominator is the value of the place of the last number of the decimal.

EXAMPLE 1:   Find the common fraction equivalent for 0.25.

The numerator of the common fraction is 25. Since the 5 is in the hundredths, 100 is the denominator.

$$\frac{25}{100} = \frac{1}{4}$$

EXAMPLE 2:   Find the decimal equivalent for $\frac{1}{2}$.

Divide the denominator into the numerator.

$$\begin{array}{r} .5 \\ 2\,)\overline{1.0} \\ \underline{1\,0} \\ 0 \end{array}$$

With decimal fractions, a 0 is added to the left of the decimal point since there was nothing there. The decimal equivalent of $\frac{1}{2}$ is 0.5.

Many common fractions do not have exact decimal equivalents. When that happens, the decimal starts to repeat itself. If it begins repeating, you may want to carry it out just a couple of places and round it off, unless otherwise instructed. Be certain you understand how the number is to be rounded off. There are two terms used in rounding off. One is to round off to decimal places. This means round off to that many places to the right of the decimal point. The other way is to round off to significant numbers. This means the numbers other than place keepers.

EXAMPLE 3:   A number such as 0.053 is a number to three *decimal places*, but the number has only two *significant numbers*. The 0 to the right of the decimal point is considered a place keeper. Be extra careful, because a number such as 0.503 has three decimal places and *three* significant numbers! Here the 0 is not just keeping a place. So the place keepers are all 0s from the decimal point up to the first non-0 number.

With this in mind, the decimal equivalent of $\frac{1}{3}$ is 0.33 and the decimal equivalent of $\frac{1}{6}$ is 0.167 when rounded off to the second repeating number.

Remember and use the following guidelines when finding fraction equivalents:

- To find the decimal equivalent, divide the denominator into the numerator. You may have to round off.
- To find the common fraction equivalent, use the value of the last number as the denominator.

Practical Problems:

• Apply the principles of decimal and common fraction equivalents to the problems in this unit.

In problems 1–5, find the decimal equivalent.

1. $\dfrac{5}{8}$ 2. $\dfrac{7}{16}$ 3. $\dfrac{4}{5}$ 4. $\dfrac{9}{32}$ 5. $\dfrac{3}{20}$

For problems 6–10, make a fraction and reduce to lowest terms.

6. 0.7   7. 0.2   8. 0.25   9. 0.375   10. 0.65

For problems 11–15, find the decimal equivalent (rounded off to 3 decimal places).

11. $\dfrac{2}{3}$ 12. $\dfrac{2}{7}$ 13. $\dfrac{1}{9}$ 14. $\dfrac{4}{15}$ 15. $\dfrac{7}{11}$

16. Technician Parker earned $229.20 for one day's work. He spent $\frac{3}{4}$ of his day repairing a wind turbine. What should be the charge for Parker's work on the wind turbine?   _____

17. A low-head hydro generator uses a flow of 2,217.6 gpm. The stream's flow has $\frac{3}{11}$ more flow than is used by the generator. How much is this extra flow?   _____

18. What is the inside diameter of this Schedule 80 PVC pressure pipe used in a geothermal well system?   _____

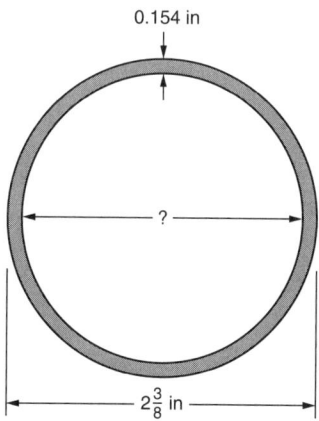

0.154 in

?

$2\frac{3}{8}$ in

19. A geothermal well is $438\frac{3}{4}$ ft deep. Piping extends 0.95 of the depth. How long is the piping?   _____

20. A PV system is producing 1,985.75 W from 8 cells. $\frac{2}{3}$ of the power is coming from 4 cells. How much power are these 4 cells producing? Round off to two decimal places.   _____

21. Recharging the heat exchanger of a geothermal system used ½ of the refrigerant in a cylinder. At the start of this job, the cylinder held 24.8 lbs of refrigerant. How much refrigerant was used?   _____

22. During one day a low-head hydro generation system produced 7,128 kWh, which was $\frac{4}{5}$ of the power used on the farm. What power did the farm use?   _____

23. Installing a geothermal heating/cooling system reduced a home's power usage to 13,667.24 kWh. If the system had used 16,873.25 kWh, what decimal portion of the power has been saved?   _____

24. A solar hot water heating system generated 5,308.44 Btu in 4 days. At this rate, how much would be generated in one week?   _____

25. Technician Brett has dug and buried the cable shown while installing a wind turbine. He has completed $\frac{3}{5}$ of the job. How much further does he have to go to complete the job?   _____

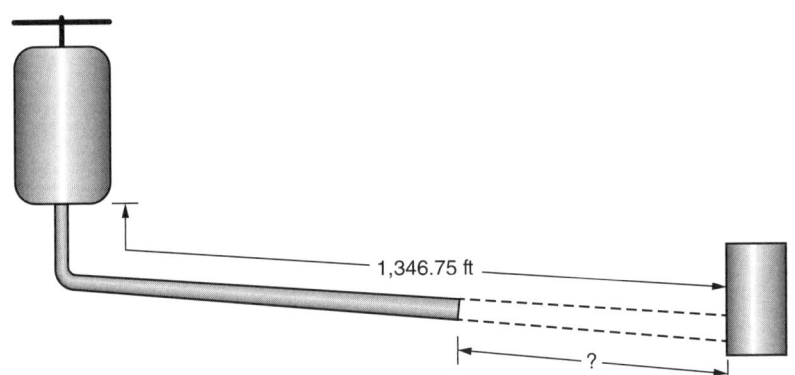

1,346.75 ft

?

UNIT 16

Combined Operations with Decimal Fractions

Basic Principles of Combined Operations with Decimal Fractions

- Review and apply the principles of addition, subtraction, multiplication, and division of decimal fractions to these problems.
- Read each problem. Decide which operation must be performed to solve the problem. (The same hints given in Unit 5 can be used here to help determine which operation to perform.)
- Perform the operation. Some problems may require the application of more than one type of operation to solve the problem. As a result, it may take more than one step to find the answer.

EXAMPLE 1: Two identical wind turbines on Nell's Wind Farm were monitored. The readings for the turbines were taken 24.5 hours apart. The two generated a total of 11,025 kWh. How much did one generator produce in one hour?

This is a two-step process. First we need to find out how much was produced by one generator. Since there were two turbines, take $\frac{1}{2}$ of the total to find the output of one turbine.

Estimating:

$$\frac{1}{2} \times 10,000 \text{ kWh} = 5,000 \text{ kWh}$$

Then divide that quantity by the number of hours to get the output for one hour.

$$5,000 \text{ kWh} \div 20 \text{ hr} = 250 \text{ kW}$$

Working the problem:

$$\frac{1}{2} \times 11{,}025 \text{ kWh} = 5{,}512\frac{1}{2} \text{ kWh} = 5{,}512.5 \text{ kWh}$$

$$5{,}512.5 \text{ kWh} \div 24.5 \text{ hr} = 225 \text{ kW}$$

Practical Problems:

In problems 1 - 12, perform the indicated operation.

1. $\begin{array}{r} 527.41 \\ 306.88 \\ +\ \ 46.07 \\ \hline \end{array}$

2. $\begin{array}{r} 64.037 \\ -\ 35.36 \\ \hline \end{array}$

3. $\begin{array}{r} 0.0758 \\ \times\ 63.94 \\ \hline \end{array}$

4. $73.41\overline{)3.868707}$

5. $49.57 + 3.864 =$ _____

6. $54.81 - 9.946 =$ _____

7. $2.307 \times 0.095 =$ _____

8. $0.0029594 \div 0.0614 =$ _____

9. $2.306 \text{ gpm} + 0.957 \text{ gpm} + 11.047 \text{ gpm} =$ _____

10. $14.818 \text{ hours} - 5.831 \text{ hours} =$ _____

11. $8.207 \text{ sq ft} \times 0.919 \text{ ft} =$ _____

12. $60.760926 \text{ MW} \div 16.413 =$ _____

13. Convert $\frac{1}{3}$ to a decimal. (Round off to 3 decimal places.) _____

14. Convert 0.625 to a fraction. (Reduce to lowest terms.) _____

15. A $9\frac{7}{8}$ ft section was cut from the conduit shown. How much conduit is left?

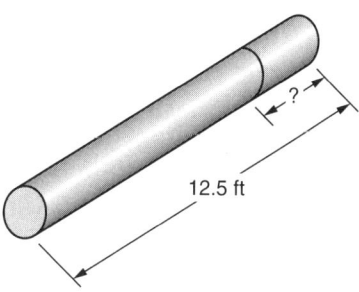

12.5 ft

16. A ditcher is rented to install cables to wind turbines. It rents for $97.35 per hour and was used for 27.6 hours. How much did it cost to rent the ditcher?

17. Three PV solar arrays produced 175.98 kW, 172.44 kW, and 179.07 kW one week. What was the total power produced by these arrays?

18. The Federal Aviation Administration (FAA) requires flashing lights on objects that exceed 200 ft above ground level. Does the wind turbine shown require flashing lights?

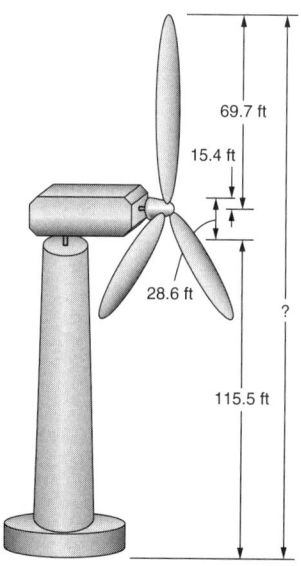

69.7 ft

15.4 ft

28.6 ft

115.5 ft

?

19. A building complex installed solar collectors on the roof of each building for each apartment. One month each of the 24 apartments saved 360.73 kWh as a result. What was the total energy saved that month? _____

20. A low-head hydro generator produces 819.84 kWh, but uses 25.27 kWh to run lights and meters at the production site. What is the net generator output? _____

21. One bag of grout, when mixed, will fill a volume of 6.38 cu ft. How many bags of grout will be need when filling a geothermal well of volume 1,207.96 cu ft? (Round off to the next higher bag.) _____

22. At a non-movable solar collector, due to the angles of the sun, the energy striking the surface in the winter is $\frac{3}{4}$ that which strikes the surface in the summer. If the summer brings 298.7 Btu/ft^2 on the surface, what is the intensity of the winter sun? _____

23. Technician Carson spent 11.25 hours on a repair job. He spent $\frac{1}{8}$ of that time driving to and from the job site. How much of the time was spent actually repairing the system? _____

24. What volume of concrete is needed to create a $7\frac{1}{2}$ ft deep foundation for a wind turbine tower with surface dimensions shown? _____

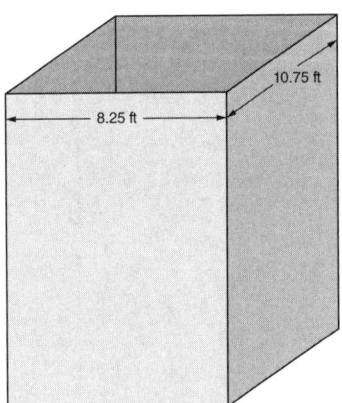

10.75 ft

8.25 ft

25. Home owner Suzanne has 2 PV panels on her roof. They produce 0.495 kW of power. She is thinking about adding 7 more. What would be the total production after completing the installation? _____

SECTION 4

Averages, Ratio, and Proportion

UNIT 17

Averages

Basic Principles of Averages

- Review denominate numbers in Section I of the Appendix.

There are times when a quantity is needed that does not have a steady value. We need a single value for a number that does not have a single value. An example of this might be the speed of the wind today. The wind speed varies over the day, so what value would you give?

One way to give a meaningful number is to take a series of readings and then determine one value that would give the same results as all of these readings taken together. This one value is called the average. An average is determined by adding the quantities and dividing by the number of inputs.

The quantities being added must have the same units. The average value may be none of the input values. The average can be determined as a whole number, a fraction, or a decimal value.

EXAMPLE 1: What is the average number of days in a month?

Estimating the answer:

12 months have about 400 days, which gives

400 days ÷ 12 = 33.3 days.

Solving the problem:

Adding the days in the 12 months gives 365 days.

365 days ÷ 12 = $30\frac{5}{12}$ days

Remember and use the following guidelines when finding averages:

- The quantities must have the same units.
- The average may be none of the input values.

Practical Problems:

- Apply the principles of averages to the problems in this unit.

If needed, round answer to 2 decimal places.

In problems 1–10, find the average value.

1. 24, 34

2. 157, 218, 194, 203

3. Round to the nearer whole number: 413, 909, 556, 843, 176, 812, 428, 621, 688

4. $15\frac{2}{7}$, $22\frac{4}{5}$, $7\frac{1}{2}$, $28\frac{7}{10}$, 19

5. 83.4, 271.3, 195.6, 248.5

6. 13 hr, 11 hr, 12 hr

7. Round to 1 decimal figure: 8 ft, 19 ft, 3 ft, 21 ft, 18 ft, 7 ft

8. 14.5 in, 31.9 in, 18.7 in

9. $1,134.25, $265.13, $2,477.68, $599.30

10. $5\frac{1}{3}$ lb, $8\frac{2}{3}$ lb, $9\frac{4}{5}$ lb, $8\frac{7}{8}$ lb

11. Four wind turbines were being monitored on Nell's Wind Farm. On one day, they produced 6,300 kWh, 5,915 kWh, 6,128 kWh, and 6,233 kWh. What was the average output per turbine?

12. While checking out the possibility of a PV installation, a home owner kept track of the number of days with at least 6 hours of sunshine during 3 months. The numbers were 22, 25, and 19. What is the average number of sunny days in a month for this period?   _____

13. A stream is measured once a day for flow for a possible low-head generator. During one week flows of 2,523 gpm, 2,612 gpm, 2,477 gpm, 2,597 gpm, 2,463 gpm, 2,605 gpm, and 2,516 gpm were recorded. What is the average flow for that week? (Round off to the nearer gpm.)   _____

14. Four geothermal wells were drilled in one housing development. The depths are 523 ft, 496 ft, 509 ft, and 517 ft. Find the average well depth in that development.   _____

15. Technician Parker worked the following overtime hours one week:

Monday – $2\frac{1}{4}$ hours,
Tuesday – 0 hours,
Wednesday – $1\frac{1}{2}$ hours,
Thursday – $2\frac{1}{3}$ hours,
Friday – $1\frac{3}{4}$ hours, and
Saturday – 4 hours.

What is Parker's average daily overtime?   _____

16. During one month, technician Makena put gas into her truck five times in four weeks. The charges for the gas were $62.45, $66.20, $58.95, $60.53, and $63.30. What is the average *weekly* charge for gas for Makena's truck?   _____

17. Manager Suzanne keeps track of the availability of the eight wind turbines on the farm she manages. During one month, the turbines were available 29 days, 29 days, 30 days, 28 days, 30 days, 30 days, 29 days, and 26 days. What was the average availability for these turbines? Round your answer to two decimal places.   _____

18. An anemometer is used to measure wind speeds. A reading was taken every 4 hours one day. The values recorded were 6.3 mph, 5.9 mph, 8.7 mph, 11.6 mph, 5.5 mph, and 6.7 mph. What was the average of these readings rounded off to the nearer tenth mph?   _____

19. The rate a well driller can drill depends upon whether it is drilling through dirt or rock and the type of rock. One rig drilled four geothermal wells at rates of 63 ft/hr, 88 ft/hr, 76 ft/hr, and 85 ft/hr. What was the average rate of drilling for these wells? _____

20. During the month of January, Mytown experiences daylight lasting 10 hr 10 min each day on average for the first week, 10 hr 22 min for the second week, 10 hr 36 min for the third, and 10 hr 52 min for the fourth. What is the daily average of daylight for the entire month? Round answer to the nearer minute. _____

21. Technician Carson kept track of repair expenses for his truck for four months. The expenses totaled $115.75 for May, $76.99 for June, $53.50 for July, and $68.84 for August. What was the average monthly expense for Carson's truck for this period? _____

22. During a two-day trade show, the Real Green Company got orders for solar hot water systems with values of $9,325.00, $9,625.00, $11,125.00, $10,875.00, $10,250.00, $9,915.00, and $10,550.
   A. What is the average cost of a system? Round answer to the nearer cent. _____
   B. What is the daily average intake? _____

23. Technician Brett has installed four PV rooftop systems in the last year. The times spent on the installations were 34 hrs, 42 hrs, 38 hrs, and 39 hrs. What was the average installation time? _____

24. One winter day has the sun's intensity measured at 1,950 Btu/m^2/hr at hour 1, 2,015 Btu/m^2/hr at hour 4, and 1,875 Btu/m^2/hr at hour 8.
   A. What is the average intensity? Round to the nearer whole number. _____
   B. What is the total 8-hour intensity? _____

25. Low-head hydro systems have been installed in an area for $5,350.75, $6,574.50, $4,935.50, and $6,225.25. What is the average cost of an installation? _____

UNIT 18

Ratio

Basic Principles of Ratio

•   Review denominate numbers in Section I of the Appendix.

Ratios are a way of comparing two numbers. The ratio can be written a number of ways. As a mathematical expression, a ratio is written as two numbers separated by a colon (:). As a statement, a ratio is stated as a ratio of one number **to** a second number. As a fraction, the first number is the numerator of the fraction and the second number (the number after the word **to**) is the denominator.

In many cases, the ratio is written as a comparison of two whole numbers. If the numbers being compared include a fraction or decimal, equivalent numbers are found that are whole numbers.

EXAMPLE 1:   Writing a ratio of $3\frac{3}{4}$ to 5 in most cases would be accomplished by first writing the ratio as $\frac{15}{4}$ to 5 and then multiplying both numbers by 4. This results in the ratio as being 15 to 20. This could then be simplified by dividing both numbers by 5 making the ratio 3 to 4.

EXAMPLE 2:   When working on a job, technician Carson drives 1.5 hours to get to the job site, then puts in 9 hours of work at the site. What is the ratio of working hours to driving hours?

The ratio would be 9 to 1.5, but the second number is not a whole number, so multiplying by 10 gives 90 to 15. This ratio can also be simplified (reduced) by dividing by 15, resulting in a ratio of 6 to 1.

There are times when ratios are left having one number be a decimal number. Almost always in these cases, the other number is 1. An example of this is a gear ratio where one shaft spins, for example, 1.4 times for each rotation of the other shaft. The ratio would be left as 1.4:1. If the second number is a number other than 1, you would divide both numbers by that second number to get the second number to be equal to 1.

Remember when working with ratios:

- The quantities are usually expressed in whole numbers.
- The ratios are usually reduced to as low whole numbers as possible.

Practical Problems:

- Apply the principles of ratio to the problems in this unit.

In problems 1–5, find the ratio as whole numbers.

1. 25:35 _____

2. 28:7 _____

3. $\dfrac{7}{9} : \dfrac{5}{9}$ _____

4. 14.6:8.2 _____

5. $\dfrac{2}{7} : \dfrac{6}{11}$ _____

In problems 6–9, find the ratio with the second number being 1.

6. 36:15 _____

7. 2:5 _____

8. $\dfrac{3}{5} : \dfrac{4}{5}$ _____

9. 14.7:43.4 _____

Use the lengths of the following line segments to determine the ratios indicated in problems 10–17.

A ├───────────────────────────┤

B ├──────────┤

C ├────────────────────────────────────┤

D ├────────────────────────────┤

10. A to B   _____

11. A to C   _____

12. B:C   _____

13. D:B   _____

14. D to A   _____

15. C to D   _____

16. C:B   _____

17. A to D   _____

18. Grout for a geothermal system well is made using 350 lbs of silica sand to 50 lbs of grout mix. What is the ratio of sand to grout mix for this well?   _____

19. A small wind turbine generator produces 300 W of power in a 10 m/s wind. What is the ratio of power to wind speed for this turbine generator?   _____

20. During the month of January, Mytown receives 94.5 W from the sun for each square meter of surface. Covering that square meter with PV cells produces 18.9 W of electrical power. Find the ratio of solar input to electrical output for this town. Round off to two decimal places.   _____

21. A home owner has 1,375 ft of well dug for a 2½ ton heat pump geothermal heat exchanger. Find the ratio of feet of well to ton of heat pump capacity.

_____

22. A wind turbine with 66 ft blades produces 225 kW of power. What is the ratio of turbine power to ft of blade?

_____

23. A low-head hydro system produces 8 kW of power using a flow of 2,536 gpm. Determine the ratio of kW to gpm.

_____

24. A geothermal well 450 ft deep has a volume of 942.5 cu ft. What is the ratio of volume to depth for this well?

_____

25. A wind turbine with a blade length of 33 ft has a turbine footprint (area restricted to other turbines) of 18,816.55 sq ft. What is the ratio of sq ft footprint per foot of blade length?

_____

UNIT 19

Proportion

Basic Principles of Proportion

- Review denominate numbers in Section I of the Appendix.

Ratios are relationships between two quantities. There are times where there are two or more situations where we want the relationships to be the same, in other words, two situations where we have the same ratio. In these cases we can set the two ratios equal to each other, since they are the same ratio. This is called a proportion. A proportion is two ratios equal to each other. Care must be taken to have the two ratios in the same order, so that one is not the inverse of the other.

A proportion can be written as $A{:}B = C{:}D$ or $\frac{A}{B} = \frac{C}{D}$. When working with proportions, usually, one quantity is not known. The proportion is used to find that unknown value. It is found by solving the equation $\frac{A}{B} = \frac{C}{D}$ for the missing quantity. The equation is solved by cross multiplying, which gives

$$A \times D = B \times C$$

Next, arrange the term that you are solving for on the left side of the equal sign. If the unknown is either A or D, keep the equation as it is. If the unknown is B or C, then swap the two sides of the equal sign.

$$B \times C = A \times D$$

As an example, solve for C. We use the equation just written. We want to get C on the left side of the equal sign by itself. We do that by dividing both sides by B. This leaves us with C on the left side.

$$C = \frac{A \times D}{B}$$

Substitute numbers for the letters and perform the operations.

EXAMPLE 1:   Solve the proportion: $3{:}7 = x{:}28$.

This can be written as: $\frac{3}{7} = \frac{x}{28}$

$$x \times 7 = 3 \times 28$$

$$x = \frac{3 \times 28}{7}$$

Estimating the answer gives $\dfrac{3 \times 30}{7} = \dfrac{90}{7} = 12\dfrac{6}{7}$ or about 13.

Solving the problem $x = \dfrac{3 \times \overset{4}{\cancel{28}}}{\underset{1}{\cancel{7}}}$

$$x = 3 \times 4 = 12$$

One application for proportions is with triangles. If two triangles have the same size angles, the lengths of their corresponding sides will be proportional. This only works if the angles for the two triangles are the same. Two such triangles are called similar triangles. Corresponding sides are the sides that have the same location with regard to the same angles.

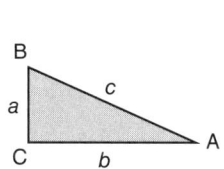
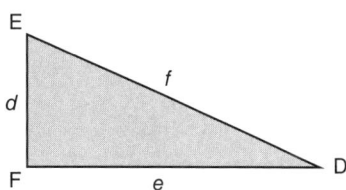

In these triangles, angles B and E are the same size, angles C and F are the same, and angles A and D are the same. What this means is that the ratio of lengths of sides $a{:}d$ is the same as the ratio of lengths of sides $b{:}e$ and also sides $c{:}f$. A proportion can be set up.

$$a{:}d = b{:}e = c{:}f$$

What is important is that the angles are the same size, not the orientation of the triangle. Here are the same triangles, which would have the same proportions, but may not look it.

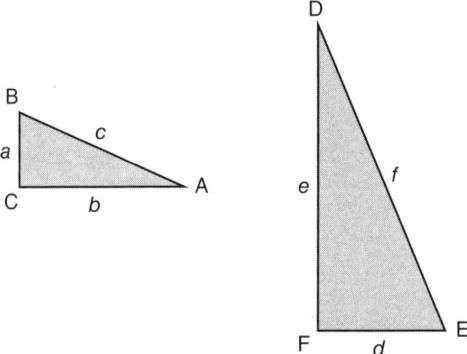

Any two sets of corresponding sides will have the same proportion.

EXAMPLE 2: Find the length of side e given $a = 1$, $b = 4$, and $d = 2$.

Solving:

$$\frac{a}{d} = \frac{b}{e}$$

$$\frac{1}{2} = \frac{4}{e}$$

Cross multiplying and dividing gives

$$e = \frac{4 \times 2}{1} = 8$$

Remember when working with ratios:

- The ratios must be in the same order when setting them equal to each other.

Practical Problems

- Apply the principles of proportion to the problems in this unit.

In problems 1–10, find the missing value.

1. $\dfrac{6}{11} = \dfrac{x}{44}$ _____

2. $\dfrac{15}{2} = \dfrac{45}{x}$ _____

3. $\dfrac{8}{5} = \dfrac{x}{12}$ _____

4. $\dfrac{3}{4} = \dfrac{17}{x}$ _____

5. $\dfrac{3/8}{1} = \dfrac{11/8}{x}$ _____

6. $\dfrac{4.3}{6.2} = \dfrac{x}{31}$ _____

7. $5:9 = x:81$ _____

8. $7:2 = 49:?$ _____

9. $3.14:5 = ?:8.5$ _____

10. $\dfrac{1}{16}:2 = 3\dfrac{1}{4}:?$ _____

11. Find the lengths of sides x and y.  _____

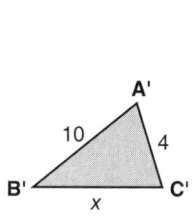
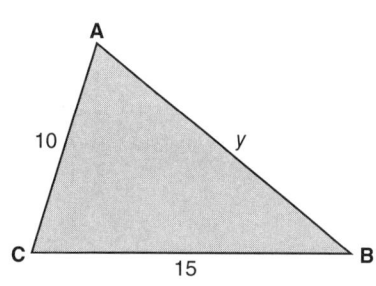

12. Grout used in a geothermal well is made mixing 21 gallons of water with 400 lbs of sand and grout mix. For the same properties for the grout, how many gallons of water are needed for 1,000 lbs of sand and grout mix?

13. A half-ton heating/cooling system requires 275 feet of piping for a horizontal geothermal loop heat exchanger. A house in the same area requires a $3\frac{1}{2}$-ton system. How much piping will this system require for its ground loop?

14. In order to properly support a PV panel at the correct angle, what is the correct length of the tall support?

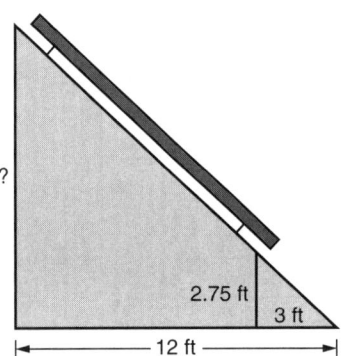

15. A wind turbine produces 1,050 W in a wind of speed 8 m/s. What power would be produced in a wind of 10.5 m/s?

16. A low-head hydro system has a straight penstock that drops 14 ft in height as it travels 25 ft towards the turbine. How far does the water drop for its entire 90 ft travel?

17. Three PV panels have a weight of 102 lbs. What would be the weight of 16 panels?

18. A guy wire supporting a wind turbine tower is attached 15 ft above the ground and anchored to the ground at a point 9 ft from the tower. How far from the tower would an anchor be to have the same angle for a second guy wire when attached to the tower 25 ft above the ground?

19. Wind turbines are arranged in a straight line. Each of these turbines has a blade length of 49.5 ft and is set 306.9 ft apart from the next turbine. How far apart should turbines with blade lengths of 56.1 ft be placed to keep the same ratio?

_____

20. The price of #4/0 wire depends upon the amount purchased. It costs $438.75 for 75 ft of wire. At that same rate, how much would 400 ft cost?

_____

21. A solar panel with an area of 32 sq ft has a coolant flow of 0.8 gpm. What flow should be measured for an 80 sq ft collector?

_____

22. It is recommended that a $1\frac{1}{2}$-ton heating/cooling system have $787\frac{1}{2}$ ft of well as the heat sink. How many ft of well are needed for a $2\frac{1}{2}$-ton system?

_____

23. A PV system produces 2.246 kWh of electrical energy on an average winter day of 3.37 hours of sunlight. How much energy would be produced on a typical summer day of 4.69 hours of sunlight? Round off to two decimal places.

_____

24. Four wind turbines are producing 1,536 W of power at a rural location. How much would seven similar wind turbines produce?

_____

25. It took technician Makena 19.5 hours to install 3 solar collector panels. At that rate, how long will it take to install 11 panels?

_____

SECTION

Percentage, Discount, Markup, and Efficiency

UNIT 20

Percentage

Basic Principles of Percentage

- Review denominate numbers in Section I of the Appendix.

Percentage is a measure of parts of a whole based on hundredths. Percent is a way of writing decimal numbers as whole numbers. Percents are identified with a percent sign (%). Decimal numbers are written as percents by moving the decimal point *two* places to the right of its normal position and adding a percent sign.

EXAMPLE 1: Write 0.25 as a percent.

Move the decimal point two places to the right, making the number 25, and add a percent sign, making the answer 25%.

Instead of having a decimal number 0.25, you have a whole number 25%.

If the original number was a mixed decimal number, the decimal point is still moved two places to the right, so 2.4 becomes 240%. A small number may become a decimal percent; 0.034 becomes 3.4%.

When working with percents in mathematics, the number must be written as a decimal number, so 17% must be written by moving the decimal point two places to the left and dropping the % sign: 17% = 0.17. It might be necessary to add 0s to some of the numbers. For example, 3% becomes 0.03.

When determining the percent, form a ratio of $\dfrac{final\ amount}{base\ amount}$. Caution: There are times when the final amount is larger than the base amount. (When that happens, the percent is larger than 100%.) Most of the time the ratio can be written as $\frac{part}{whole}$. In either case, determine the decimal equivalent for the ratio, then move the decimal point two places to the right and add the percent sign.

Most percentages can be written in the form:

Some % **of** a number (*base amount*) **is** another number (*final amount*)

This can be written as a mathematical equation by replacing the **of** with a times sign ($\times$) and the **is** with an $=$ sign. Rearranging the equation results in

$$\% = \frac{final\ amount}{base\ amount}$$

This can be set up as a proportion:

$$\frac{actual\ \%}{100\%} = \frac{final\ amount}{base\ amount}$$

Change the percent to a decimal

$$\frac{decimal(from\%)}{1} = \frac{final\ amount}{base\ amount}$$

Like any proportion, any of the quantities can be found knowing the other three. The quantities given in a problem can be identified by thinking of the problem in terms of the general form. The quantities can then be substituted into the proportion and solved.

EXAMPLE 2: Find 15% of 20.

15 is the percent since it has the sign. It needs to be converted to a decimal. 15% = 0.15. 20 is the base amount since it is immediately after **of**. So the problem becomes

$$\frac{0.15}{1} = \frac{?}{20}$$

$$.15 \times 20 = ? \times 1$$

$$? = 3$$

EXAMPLE 3: 8 is 20% of what number?

This can be put into the general form by rearranging it to read:

20% of what number is 8?

Now we see that 20% = 0.2. The unknown is the base amount, and 8 is the final amount.

$$\frac{0.2}{1} = \frac{8}{?}$$

$$.2 \times ? = 8 \times 1$$

$$? = \frac{8}{.2}$$

$$= 40$$

EXAMPLE 4: 25 is what % of 20?

Rearranging this like the previous example:

What % of 20 is 25?

20 is the base amount, and 25 is the final amount.

$$\frac{?}{1} = \frac{25}{20}$$

$$? = 1.25$$

This makes 125% as the percentage.

Remember when working with percentages:

- The percent must be converted to a decimal to be used in the problem.
- Putting it in general form makes it easier to identify which number is which.

Practical Problems:

• Apply the principles of percentage to the problems in this unit.

In problems 1–4, convert to a decimal.

1. 35%   _____

2. 17.6%   _____

3. 2½%   _____

4. 120%   _____

In problems 5–8, convert to percent.

5. 0.41   _____

6. 0.7   _____

7. 1.5   _____

8. 0.03   _____

9. What is 4% of 17?   _____

10. Find 250% of 112.   _____

11. 42 is 16% of what number?   _____

12. 6 is 6% of what number?   _____

13. 23 is what % of 542? Round off to one decimal place.   _____

14. 74 is what % of 20?   _____

15. One day in early spring has 9 hours of sunlight. What % of that day has sunlight?   _____

16. A stream has a flow of 2,500 gpm. A low-head hydro system on that stream uses 2,200 gpm. What % of the stream is used by the hydro system?   _____

17. A wind turbine, which produces 734 MWh in a year at one wind speed, produces only 128 MWh at a slower wind speed. What % of the faster production is the slower production? Round off to one decimal place.   _____

18. The total hourly sunlight is delivering 5.2 kW per hour on a solar collector. 74% is converted to usable heat by the collector. How much usable heat will the collector put out?   _____

19. Technician Parker wrote a bill for a repair job. He listed the labor cost as $120.00. This was 48% of the entire bill. What was the total of the entire repair bill?   _____

20. An individual wind turbine produces 26 kWh per day. This is 8% of the house's daily electrical needs. How many kWh per day does the house use?   _____

21. Installing a PV electrical system requires 32 man-hours. Technician Hayden has already worked 20 man-hours. What % of the job has been completed?   _____

22. 22 gallons of water are used for a batch of geothermal well grout. This is 53% of the batch volume. What is the volume of 1 batch of grout? Round off to one decimal place.   _____

23. Mytown averages 3.37 hours of sunlight each day during the winter. It also averages 4.88 hours of sunlight during the summer. What percent of the summer hours is the winter hours? Round off to the nearer percent.   _____

24. Determine the % of the roof covered by PV panels shown. Round off to the nearer percent.   _____

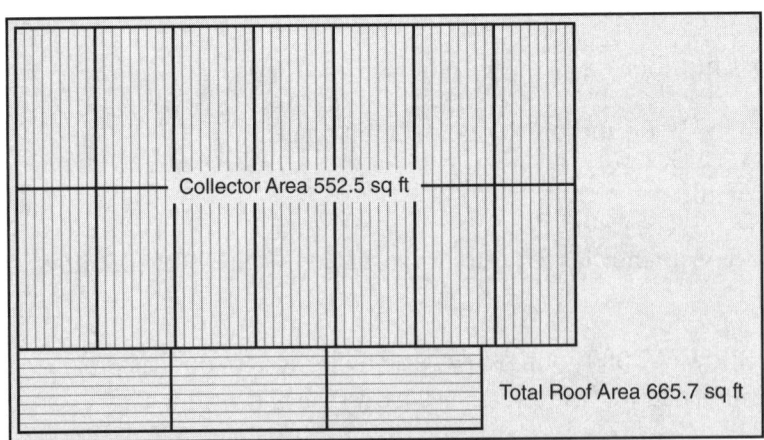

Collector Area 552.5 sq ft

Total Roof Area 665.7 sq ft

25. Nell's Wind Farm added 360 kW of turbine capacity. That represents 15% of the total capacity. What is the capacity of the farm?   _____

Discounts and Markups

Basic Principles of Discounts and Markups

Discounts and markups are basically percent problems. They deal with money, so all of the values will either be percents or decimals rounded to the nearer hundredth (cent). When working discount or markup problems, first multiply the percent times the original price. Next, if working a **discount** problem, *subtract* this value from the original price. This gives the discounted price. If working a **markup** problem, *add* this value to the original price. This gives the marked up price.

EXAMPLE 1: A bill of $140.00 is given a discount of 6%. What is the discounted price?

Estimating the answer: $100.00 $\times$ 0.06 = $6.00

$100.00 $-$ 6.00 = $94.00

Solving the problem:

$140.00 $\times$ 0.06 = $8.40

$140.00 $-$ 8.40 = $131.60

EXAMPLE 2: A supply store stocks a low-head generator. It purchased one for $667.25. The store marks the generator up by 45%. What is the selling price at the store?

Estimating the answer: $700.00 $\times$ 0.5 = $350.00

$700.00 + 350.00 = $1,050.00

Solving the problem:

$$\$667.25 \times 0.45 = \$300.2625 \text{ or } \$300.26$$

$$\$667.25 + 300.26 = \$967.51$$

There are stores that have special deals for special customers. The special deal is a double discount. A double discount determines the discounted price using the first discount. This is called the first discounted price. The second discount is used with the first discounted price. Then the second discounted price is found by subtracting the second discount from the first discounted price.

EXAMPLE 3: Nell's Wind Farm bought supplies from the Big Wind Supply Store. The bill totals $273.66. The store gives Nell a double discount of 11% and 5%. What is the cost to Nell for these supplies?

Estimating the answer: $\$300.00 \times 0.1 = \$30$

$$\$300.00 - 30.00 = \$270.00 \text{ (The first discounted price)}$$

$$\$270.00 \times 0.05 = \$13.50$$

$$\$270.00 - 13.50 = \$256.50 \text{ (The second discounted price)}$$

Solving the problem:

$$\$273.66 \times 0.11 = \$30.1026 \text{ or } \$30.10$$

$$\$273.66 - 30.10 = \$243.56$$

$$\$243.56 \times 0.05 = \$12.178 \text{ or } \$12.18$$

$$\$243.56 - 12.18 = \$231.38$$

Another situation dealing with discounts occurs when some bills are sent. These bills contain a notation such as **1%10/Net 30**. This means that if the bill is paid within 10 days, there will be a 1% discount. If not paid within 10 days, the full bill is to be paid within 30 days. After 30 days, late charges can be added.

Remember when working with discounts and markups:

- Use the percent to determine the dollar amount that will be subtracted if a discount or added if a markup is determined.

• Double discounts use the first discounted price to determine the second discount.
• Round off all answers to the nearer cent.

Practical Problems:

• Apply the principles of discounts or markups to the problems in this unit.

1. What is the discount on a $2,478.75 bill if the discount is 8%?   _____

2. Determine the discounted price if a bill of $351.25 is discounted 7%.   _____

3. Find the markup if it is 23% on a value of $973.17.   _____

4. The cost of a pump to a supplier is $615.99. The store marks up the price by 35%. What is the price the supplier puts on the pump?   _____

5. Replacement parts for a wind turbine total $75.24. A 13% discount is given. What is the discounted price for the parts?   _____

6. A 500 ft coil of #4/0 wire is priced at $1,650.00. The 5-J Construction Company gets a 5.5% discount. What price would the 5-J company pay?   _____

7. A plastic replacement turbine blade for a low-head hydro generator is marked at $140.50. A discount of 9% is offered. What is the asking price for the turbine blade?   _____

8. The piping for a low-head hydro penstock is listed at $286.95. It can be purchased with a discount of 8%. What would be the savings?   _____

9. A PV panel costs the distributor $460.00. The distributor marks the panel up by 30%. What is the distributor's selling price for the panel?   _____

10. The cost of a 30 ft tower for a 250-watt wind turbine costs a dealer $4,750.00. The dealer marks up the tower cost by 27%. What price is the dealer asking for the tower?   _____

11. A building supply store offers discounts to contractors. The discounts are 6% for electrical supplies, plumbing supplies 10%, and other supplies and hardware 7%. While installing a low-head hydro system the following supplies were purchased. What is the contractor's total bill? _____

	WE HAVE IT SUPPLY COMPANY Yourtown, MD		
Qty	**ITEM**	**Price Per**	**PRICE**
200 feet	4/0 cable	$3.50/ft	
1	junction box	$5.99	
380 feet	6" PVC piping	$24.99/10'	
2	6" 90 elbow	$6.89	
2	6" 45 elbow	$5.79	
1 box	#8 x 2 Wood Screws	$7.99	
3	Caulk cartridges	$6.39	
1 box	8d - 2 1/2 Nails	16.99	
2 gal	Rust resistant paint - brown	$34.95	
		TOTAL	

12. A bill to a contractor is $734.27 with discounts of 13% and 3%. What is the amount the contractor owes? _____

13. What is the price of a bill of $2,885.56 after a double discount of 9% and 5%? _____

14. A single 245 W PV panel costs $588.00. If greater than 10 panels are purchased, the panel price is discounted by 7%. What would be the total cost of 12 panels? _____

15. The components of a solar hot water heating system cost $1,155.75. The owner of a 6-unit condo can purchase 6 systems for a price discounted by 8% each. What is the total price for the 6 systems? _____

16. A distributor ordered a low-head hydro system transformer at a cost of $723.00. The distributor marked the transformer up 31%. What is the marked up price?

_____

17. Wind turbine parts are stocked in a store. The parts were purchased for $2,587.43. They were all marked up 47%. What would be the new value of the parts?

_____

18. A garage offers a 7% discount for repair trucks worked on during the winter months. Technician Makena takes her truck in for repairs in December. The bill before the discount is $383.29. What is Makena's discounted bill?

_____

19. A contractor installing a solar hot water panel buys the panel and marks it up 27%. That is his income from the job. If the panel cost the contractor $2,750.00, what is the contractor's income?

_____

20. A do-it-yourself home owner plans to install the electrical wiring for a wind turbine himself. The place where the wind turbine was purchased will sell the home owner the supplies at a 2% discount. The bill for the supplies was $643.78. How much was the home owner charged?

_____

21. The WeGotIt Supply Co. offers a 4% discount to contractors. One contractor bought $432.47 worth of supplies before the discount is applied. If the bill is paid within 7 days, an additional 1.5% is taken off.

A. What is the discounted bill?

_____

B. How much must be paid on day 5?

_____

22. A home owner gets a bill for $307.62 for repairs to his solar hot water system. The bill states that if the bill is paid within 14 days, a 1.5% discount will be applied to the bill. If the bill is paid within 7 days, a 1% discount will be taken off the discounted price.

A. How much would the home owner pay if the bill is paid on day 12?

_____

B. How much would the home owner pay if the bill is paid on day 6?

_____

23. A contractor has an agreement with a distributor that if he pays the bill within 10 days, a 2.5% discount is applied. If paid within 20 days, a 1% discount is applied. The bill is $3,562.98.

A. What is the bill if paid on day 8?   _____

B. What is the bill if paid on day 15?   _____

24. A bill totaling $819.44 has the notation 3%10/Net30. What is the bill's amount if paid on day 5?   _____

25. A bill of $1,794.77 is marked 2%10/Net30 with a statement of a 5% late penalty.

A. How much is the bill if paid in 8 days?   _____

B. How much would the bill be if paid on day 35?   _____

UNIT 22

Efficiency

Basic Principles of Efficiency

One definition of efficiency is a measure of how well energy can be converted from one form to another. The energy conversion involves a device, and efficiency is a measure of the useful output of the device compared to the input of the device. It is usually expressed in percent. The percent is usually found by forming a ratio of the energy output to the energy input. The decimal equivalent is determined, and the result is expressed as a percent.

EXAMPLE 1: A step-up transformer takes electrical energy from a wind turbine and transforms it to the standard operating voltage of 240V. The transformer takes 300 Wh of energy from the turbine and changes it to 294 Wh of household electrical energy. Find the efficiency of the transformer.

The output of the transformer is 294 Wh and the input is 300 Wh. $\text{Efficiency} = \dfrac{\text{output}}{\text{input}}$.

Estimating the answer:

$$\text{Efficiency} = \frac{\text{output}}{\text{input}} = \frac{300 \text{ Wh}}{300 \text{ Wh}} = 1 \text{ or } 100\%$$

(Since we rounded, the actual answer should be close to 100%, but not exactly 100%.)

Solving the problem:

$$\text{Efficiency} = \frac{\text{output}}{\text{input}} = \frac{294 \text{ Wh}}{300 \text{ wh}} = 0.98 = 98\%$$

117

Whenever energy changes form, some is lost to heat or some other form of unusable energy. The efficiency will not be 100%. Some energy will be lost.

There are times when the efficiency is known, but either the input or the output is not known. The above expression can be rearranged to solve for each of those terms.

$$\text{Input} = \frac{\text{output}}{\text{efficiency}}$$

Or

$$\text{Output} = \text{input} \times \text{efficiency}$$

Heat pumps measure their efficiency by comparing the amount of heat that is moved (into the building during heating or out of the building during cooling) and comparing it to the energy used by the pumping system. This is called the coefficient of performance (COP).

$$\text{COP} = \frac{\text{Amount of heat moved in kW}}{\text{Energy useed by the pump in kW}}$$

The answer is left as a decimal fraction. Values of 3 to 4 for the COP are attainable and quite efficient. Converted to percents, these would be 300% to 400%.

Remember:

- Read each problem and determine what is needed. Use the equation that has the unknown by itself on the left side of the equal sign.
- Express efficiency in percent, but convert it to a decimal to work any problem.

Practical Problems:

- Apply the principles of efficiency to the problems in this unit.

1. Find the efficiency when 17 kW is produced from an input of 25 kW.   _____

2. An input of 117.8 Btu/h results in 106.4 Btu/h being useful. Find the efficiency to the nearer percent.   _____

3. To be at least 82% efficient, what output must be achieved from an input of 240 kW?   _____

4. What input would be needed to achieve 75% efficiency with an output of 94 Btu/h? Round to the nearer tenth.

5. Wind blowing at 15 mph has available energy of 1,354 W for a wind turbine with a 10-foot diameter rotor. The turbine produces 433 W of electrical power. What is the efficiency of that particular turbine? Round answer to the nearest whole percent.

6. A stream has a flow of 600 gpm. A dam provides 6 ft of head, allowing the flow to turn a low-head hydro generator. The available energy is 682 W. The generator output is 354.64 W. What is the efficiency of the hydro system?

7. Different solar PV panels have different efficiencies. One PV panel has 1,361 W of power shining on one square meter of it, resulting in the production of 170 W of electrical power. What is the efficiency of this PV panel? Round to the nearer tenth percent.

8. What is the efficiency of the solar panel shown?

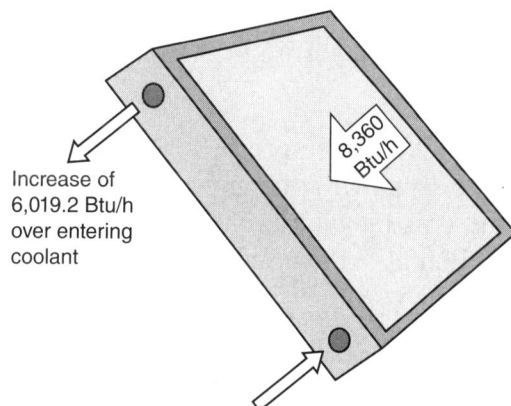

Increase of
6,019.2 Btu/h
over entering
coolant

8,360
Btu/h

9. A geothermal heating/cooling system in the heating mode delivers 12.795 Btu/h of heat for each watt (3.412 Btu/h) of electrical power used by the system. What is the efficiency of this system?

10. A low-head dam has water that has a potential energy of 840 W. What electrical energy is generated with a system that has an efficiency of 56%?

11. A 1 sq m solar heating panel has 1,250 W of power striking it, which equates to 4,265 Btu/h heat. The collector is 78% efficient. How much usable heat is removed from this collector?

_____

12. A 7 ft high dam feeds a hydro generator, producing 450 W. Using the same flow and the same efficiency, how much higher must the water level be raised to increase the output to 500 W? Round to the nearer hundredth. (Hint: The energy of the water is proportional to the height of the water.)

_____

13. With 17 mph winds, one wind turbine has an efficiency of 42%. If the wind's power at that speed is 1,405 kW, what is the output of the wind generator?

_____

14. One PV panel normally produces 270 W. On a sunny summer day, the panel gets hot and produces only 245 W. What is the efficiency of the panel compared to normal production? Round to the nearer tenth percent.

_____

15. Improvements in wind turbine blade designs over the past 10 years have increased the output of the turbine generators by 8%. If the output of the turbine had been 395 kW, what is the new output from the improved turbine blades?

_____

16. An all-electric home uses 120 kWh to produce the needed heat on a day in the winter. If the home owner changes to a heat pump, the system is 145% efficient. This means that less electrical input will produce the same output. Round to the nearer tenth.

   A. What is the new electrical input used by the new heat pump system?

_____

   B. If the system has a ground coil heat exchanger, the efficiency increases to 425% efficient. What is the input if the ground coil is used?

_____

17. Solar panels are mounted on a roof that does not point directly south. Panels pointing southwest rather than due south at the correct vertical angle are 96% efficient compared to the correct orientation. A PV panel is rated at 240 W. What would be the maximum power for a southwest-oriented panel?

_____

18. A low-head hydro system has water flowing with energy of 14,000 W (14 kW) and generates 7.9 kW electrical energy. What is the efficiency of this system? Round to the nearer percent.

_____

19. On a totally sunny day, a PV panel converts 17% of the sun's energy to electrical energy.

    A. What is the electrical output from the panel if a total of 1,200 W of solar energy fall on the panel?

_____

    B. On a cloudy day, only 120 W of solar energy fall on the panel, but the efficiency increases to 18% (due to lower panel temperatures). What is the output of the panel on a cloudy day?

_____

20. Increasing the flow of coolant through a flat-plate solar collector causes the coolant to exit at a lower temperature, but increases the efficiency of the system, since more of the heat of the plate gets removed by the coolant and not lost to ambient air cooling. A solar collector has about 2,700 Btu/hr strike it. Panel efficiency goes from 50.3% to 48.8% as the flow speed decreases from 0.54 gpm to 0.225 gpm. What is the gain in heat being captured by the system in Btu/hr at the higher flow rate as compared to the lower flow rate?

_____

21. One might consider measuring the saving-to-cost ratio as an efficiency of sorts. Installing a solar water heater system to a house cost $1,200.00. The system saves $65.00 each year. What is the "efficiency" of the system? Round to the nearer tenth percent.

_____

22. A solar panel is positioned to receive the maximum amount of solar radiation for electrical conversion during December. During May the fixed panel will be 80% as efficient (due to the angle that the sunlight will be striking the panel surface). If the panel produces 175 W during December, what will be the output during May?

_____

23. A typical Arizona town has 211 sunny days. A town in the Northeast has 93 sunny days. How much more efficient would a solar system be in Arizona than the Northeast? Round to the nearer hundredth percent.

_____

24. A heat pump uses 1.2 kW of electrical energy to remove 4.08 kW of heat energy from a building. What is the coefficient of performance for this heat pump?

                        _____

25. Find the COP of the system shown.

                        _____

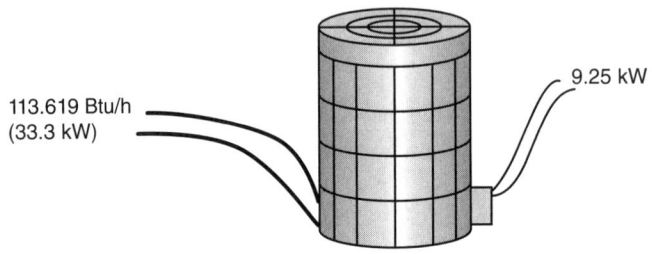

113.619 Btu/h
(33.3 kW)

9.25 kW

SECTION

Direct Measure

UNIT 23

Equivalent Units of Temperature Measure

Basic Principles of Equivalent Units of Temperature Measure

Temperature and temperature measure are important in a number of areas related to renewable energy. One of the concerns with temperature measure is that the United States uses the Fahrenheit scale to measure temperatures, while most of the rest of the world uses the Celsius scale. Measuring the temperature of an object will give two different numbers depending on which scale is being used to measure the temperature. Very often, when equipment is manufactured outside of the United States, the technical data includes Celsius temperature information. A technician should be able to correctly convert from one measuring system to the other.

A comparison of thermometers used to measure temperatures using the two different scales is shown below.

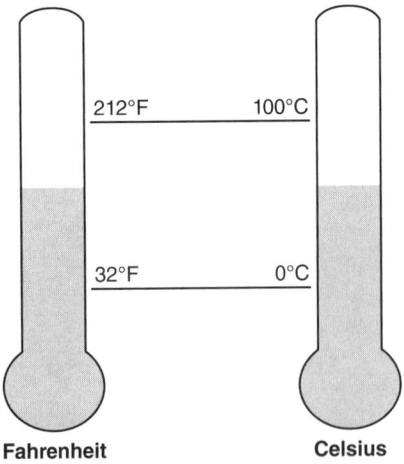

Fahrenheit   Celsius

To convert between specific temperatures on the two scales, two formulas are used. The first is to convert to a Fahrenheit temperature when a Celsius temperature is given.

$$°F = \frac{9}{5}°C + 32$$

If given a Fahrenheit temperature, conversion to a Celsius temperature is performed using

$$°C = \frac{5}{9}(°F - 32)$$

Care must be taken when using these formulas to do the correct mathematical operations in the correct order. In the first formula there are no parentheses. Make sure you multiply the °C reading by $\frac{9}{5}$ *first* and then add 32 to that number. The second formula *has* parentheses, so here subtract 32 from the °F value and *then* multiply that result by $\frac{5}{9}$. These operations *must* be done in that order.

EXAMPLE 1:   Convert 20°C to °F.

Using the first formula

$$°F = \frac{9}{5}°C + 32$$

$$= \frac{9}{5} \times 20 + 32$$

Estimating the answer: $\frac{9}{5}$ is about 2 and 32 is about 30, so

$$°F = 2 \times 20 + 30$$

$$= 40 + 30$$

$$= 70$$

Solving the problem:

$$°F = \frac{9}{5} \times 20 + 32$$

$$= 9 \times 4 + 32$$

$$= 36 + 32$$

$$= 68$$

EXAMPLE 2: Convert 59°F to °C.

Using the second formula

$$°C = \frac{5}{9}(°F - 32)$$

$$= \frac{5}{9}(59 - 32)$$

Estimating the answer: $\frac{5}{9}$ is approximately $\frac{1}{2}$, so

$$°C = \frac{1}{2}(60 - 30)$$

$$= \frac{1}{2}(30)$$

$$= 15$$

Solving the problem:

$$°C = \frac{5}{9}(59 - 32)$$

$$= \frac{5}{9}(27)$$

$$= 5 \times 3$$

$$= 15$$

Temperatures can go below 0°. When that happens, the minus sign stays with the number. When subtracting from a negative number, the number gets larger and the minus sign stays with the number. When adding to a negative number, the number will get smaller—and may go positive if it is larger than the negative number.

EXAMPLE 3: Subtract 10 from -15.

This is written $-15 - 10 = -25$.

EXAMPLE 4: Subtract 20 from 8.

This is written $8 - 20 = -12$.

EXAMPLE 5: Add 17 to -30

This is written $-30 + 17 = -13$.

EXAMPLE 6: Add 26 to -5

This is written $-5 + 26 = 21$.

There are times when the specific temperature is not needed, but the temperature difference is needed. In other words, convert the temperature difference in one scale to the temperature difference in the other scale. To do this, use the formulas

$$\text{Difference in } °F = \frac{9}{5} \times \text{Difference in } °C$$

Or

$$\text{Difference in } °C = \frac{5}{9} \times \text{Difference in } °F$$

In these cases there are no parentheses. Just multiply by the correct fraction.

EXAMPLE 7: A temperature difference of 25°C is equivalent to what temperature difference in °F?

$$\text{Difference in } °F = \frac{9}{5} \times \text{Difference in } °C$$

$$= \frac{9}{5} \times 25$$

Estimating:

$$= 2 \times 25$$

$$= 50$$

Solving the problem:

$$\text{Difference in °F} = \frac{9}{5} \times \text{Difference in °C}$$

$$\text{Difference in °F} = \frac{9}{5} \times 25$$

$$= 9 \times 5$$

$$= 45$$

Remember the following guidelines:

- When converting from Celsius to Fahrenheit, be sure to multiply first and then add.
- When converting from Fahrenheit to Celsius, be sure to subtract first and then multiply.
- When converting Celsius temperature differences to Fahrenheit temperature differences, multiply by $\frac{9}{5}$.
- When converting Fahrenheit temperature differences to Celsius temperature differences, multiply by $\frac{5}{9}$.

Practical Problems:

- Apply the principles of equivalent units of temperature measure to the problems in this unit.

Round answers to the nearer tenth degree if applicable.

Convert the following to Fahrenheit readings:

1. 65°C   _____

2. 35°C   _____

3. 7°C   _____

4. −10°C   _____

Convert the following to Celsius readings:

5. 50°F   _____

6. 135°F   _____

7. 32°F   _____

8. −10°F   _____

Convert the following temperature *differences*:

9. 45°F   _____

10. 85°F   _____

11. 30°C   _____

12. 12°C   _____

13. The inside temperature of a solar collector reaches 150°C. What is the
    Fahrenheit value for this temperature?   _____

14. A do-it-yourself person wants to build a solar collector and cover
    it with plastic that melts at 320°F. What would be the limit on the
    temperature of the collector in Celsius?   _____

15. Technician Makena measures the temperature of the liquid returning
    from a geothermal bed. The thermometer reads 6°C. What is this
    temperature on the Fahrenheit scale?   _____

16. A silicone grease was used to lubricate a wind turbine. A warning on
    the label states it should not be used in temperatures below −30°F.
    What is the Celsius equivalent for this temperature?   _____

17. A solar hot water system storage tank has water at a temperature of
    85°C. What would a Fahrenheit thermometer read for this tank? _____

18. Manager Suzanne has her house thermostat set at 72°F. What is the
    Celsius reading for this setting?   _____

19. Coolant entering a solar collector is at 60°F. What is this temperature
    in degrees Celsius?   _____

20. A bad bearing on a low-head hydro generator heats to 95°C. What is the reading on a Fahrenheit thermometer?   _____

21. Information on some PV panels indicate that if a panel's temperature increases by 20°C from its recommended temperature, panel efficiency decreases to 90%. What is this temperature increase in degrees Fahrenheit?   _____

Use the following diagram to answer the next two questions.

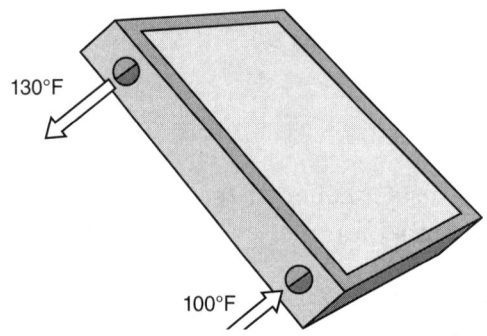

22. What is the exiting temperature in degrees Celsius?   _____

23. What is the temperature increase in the collector in degrees Celsius?   _____

24. A heat pump utilizing a geothermal well moves heat between areas with temperatures of 40°C and 15°C. What is the temperature difference in degrees Fahrenheit?   _____

25. A heat pump that does not use any geothermal cooling is used to cool a house during hot weather. It must move heat between the temperatures shown. What is the temperature difference in degrees Celsius?   _____

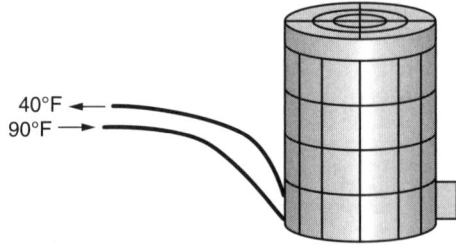

UNIT 24

Angular Measure

Basic Principles of Angular Measure

When two straight lines meet, they form an angle. The point at which the lines meet is called the apex or vertex of the angle. The two lines are known as the legs or sides of the angle. The opening between the two legs is the angle. An angle is measured using a protractor.

A protractor is a half circle (semicircle) with degree markings on it. Although protractors come in different sizes and markings, all protractors do have some common points. As a rule, all protractors have degree markings for each degree. Sometimes the markings go from 0° to 180° in each direction, while others go from 0° to 90° to 0° again.

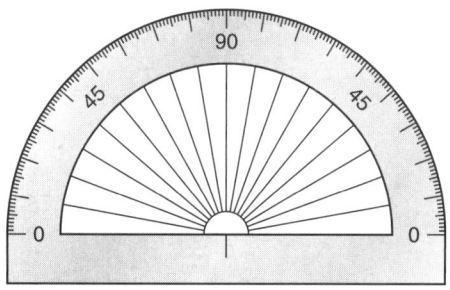

The size of an angle is measured by placing the center point of the protractor (the mark on the line between the two 0s) on the apex of the angle. Rotate the protractor so that the line to one of the 0s lies right on top of one of the sides of the angle while keeping the center point of the protractor on the apex of the angle. Next simply read off the angle where the other side of the angle crosses the protractor. If the angle is greater than 90°, care must be taken to read the correct scale or to count the number of degrees greater than 90. Many times the lengths of the sides of the angle are small. These sides can be extended before measuring the angle.

EXAMPLE 1:   Measure the following angle:

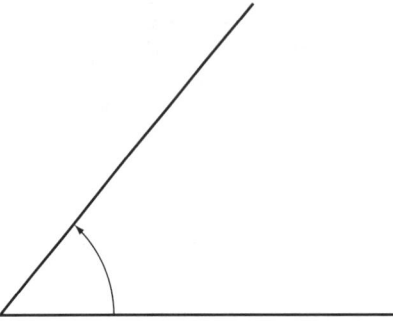

Align the 0 angle along one leg, with the vertex at the center. Read the angle where the second leg crosses the protractor. Because the 0 is on the inner set of numbers with this protractor, use the inner set of numbers to read the angle.

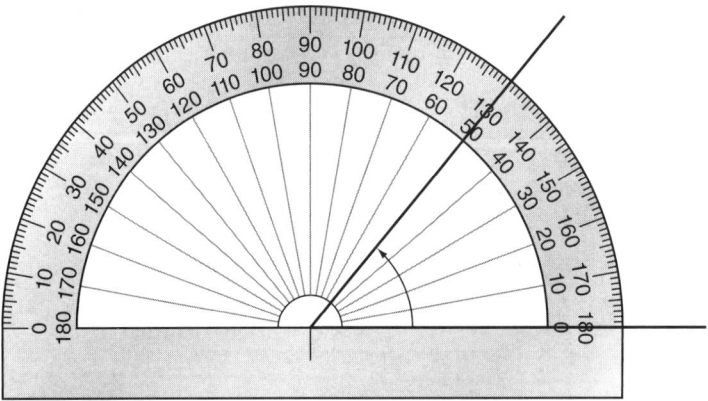

The angle is 50°.

There are some important and well-known angles. If the two legs lie on top of one another, the angle is an entire circle—which is 360°. If the two legs form a straight line, the angle is 180° and is called a semicircle. If one leg is horizontal and the other leg is vertical, the angle is 90° and is called a right angle. The orientation of the legs do not have to be horizontal and vertical; any 90° angle is called a right angle.

If the angle is greater than 180°, measure the smaller angle and then subtract that angle from 360°.

Remember: To measure an angle using a protractor, always place the center mark of the flat edge at the apex or vertex of the angle (the point where the two lines meet). Rotate the protractor around that point so that one section of the flat edge is aligned with one of the lines (sides) of the angle. The size of the angle is read where the other side of the angle crosses the curved section of the protractor. If the angle is larger than a straight line (180°), measure the smaller angle and subtract it from 360°.

There are times when angles have to be calculated. In many of those problems, reference is made to the horizon, horizontal, or to the vertical. That would help determine if you need to be concerned with 90° (vertical) or 180° (horizontal).

EXAMPLE 2: A solar panel is angled 125° from the horizontal. How far is it from the vertical?

The vertical is 90°. Since this is larger than 90°, the angle is past the vertical. The angle beyond the vertical can be found by subtracting 90° from 125°.

$$125° - 90° = 35°$$

So the panel is angled 35° beyond the vertical.

Practical Problems:

- Apply the principles of angular measurement to the problems in this unit.

For problems 1–10, measure the angles to the nearer degree.

1.

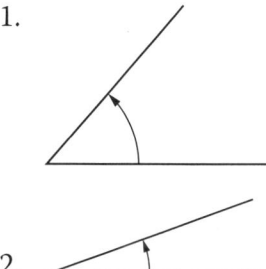

_____

2.

_____

3.

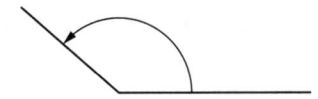

_____

4.

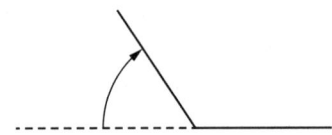

_____

5.

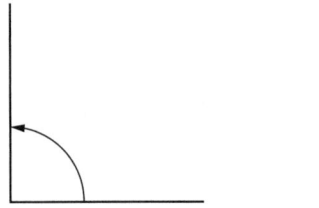

_____

6.

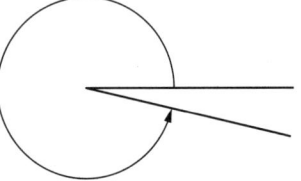

_____
_____
_____
_____

7.

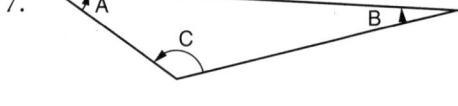

_____
_____
_____
_____

8.

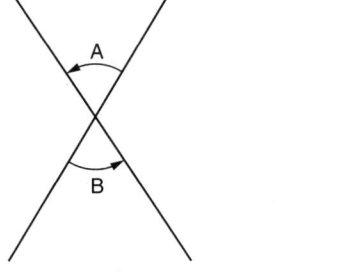

_____
_____

9.

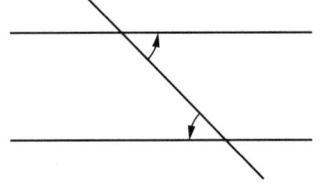

_____
_____

10.

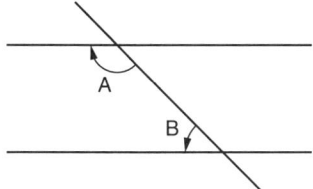

11. What is the angle of the roof shown with respect to the horizon?

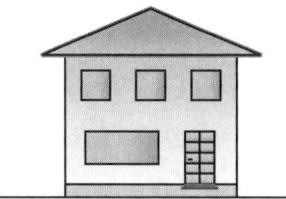

12. A penstock for a low-head hydro has a constant slope. What is its angle below the horizon?

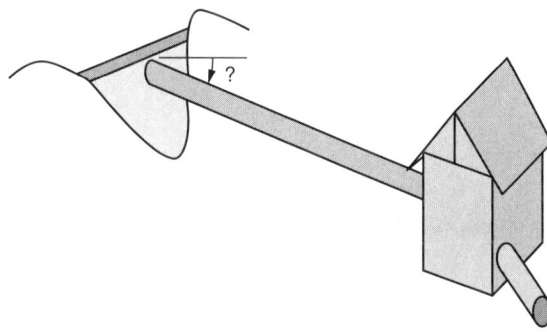

13. A solar collector is to be supported on a roof. The collector is to have an angle of 41° with the horizon. The roof has an angle of 30° with the horizon. What angle should the support brackets have that hold the solar panel to the roof?

14. A guy wire helps support a wind turbine tower as shown.

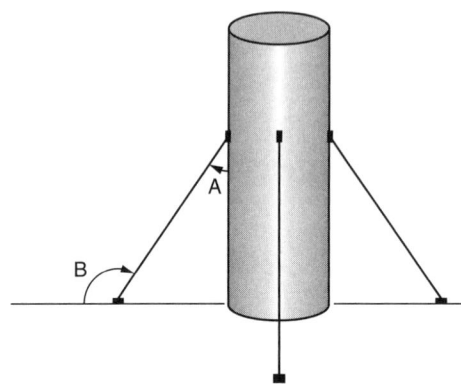

    A. What angle does the guy wire make with the tower? _____

    B. What angle does the guy wire make with the ground? _____

15. A wind turbine rotates to keep directly into the wind. One day, Nell's Wind Farm has wind varying between the directions shown. Through how big an angle do Nell's wind turbines rotate on this day? _____

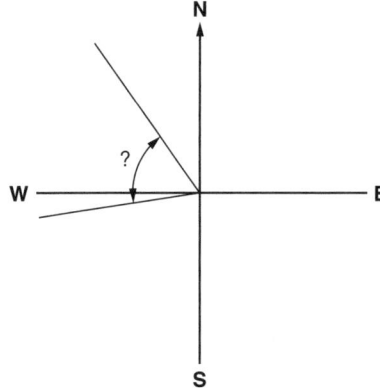

16. An extension ladder safely leans against the side of a house as shown. What angle does the ladder make with the house?   _____

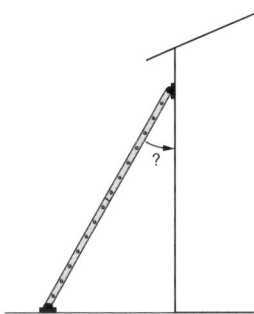

17. A penstock makes a change in the slope angle as shown. What angle must the pipe elbow have? Hint: Find the difference between angle 2 and angle 1.   _____

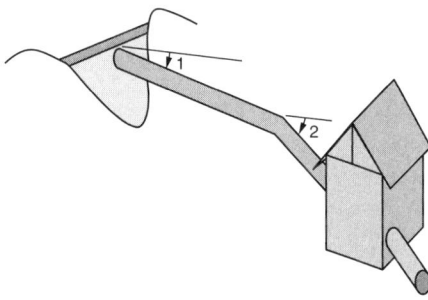

18. The PV panel is oriented as shown. What angle with the vertical does the panel make?   _____

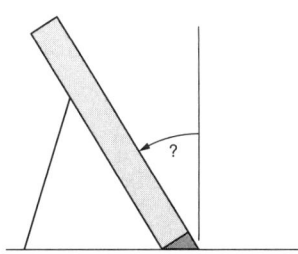

19. A wind turbine has three blades equally placed around the hub. What is the angle between any two blades? _____

20. What angle must the elbow for the solar hot water system have in the figure shown? _____

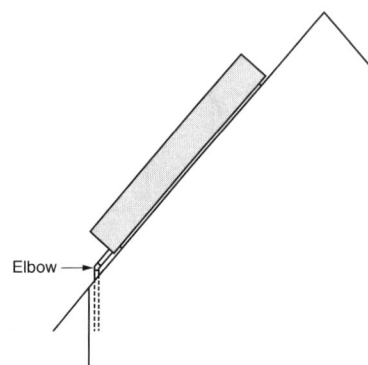

Elbow

21. A house is not oriented directly south. The owner angles a PV panel so that it is due south. What is the angle the panel makes with the house? _____

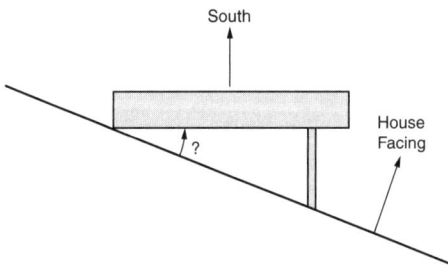

South

House Facing

?

22. What is the angle made by the support legs for the wind turbine tower shown? _____

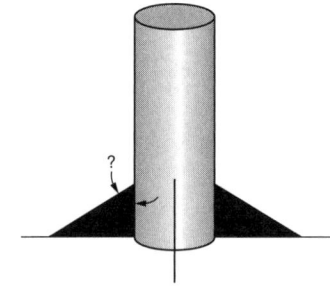

?

23. A very large wind turbine blade is not straight, but made with an angle shown. What is the angle of the blade?   _____

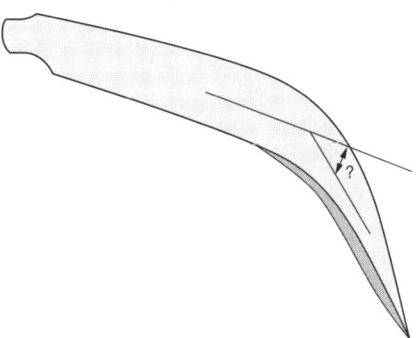

24. The piping for a geothermal horizontal bed makes the bend shown. What is the angle the piping makes?   _____

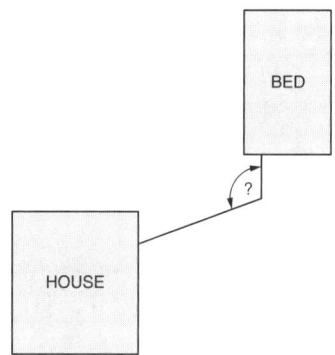

25. A penstock makes a bend as it enters the generator building. What is the angle of the bend?   _____

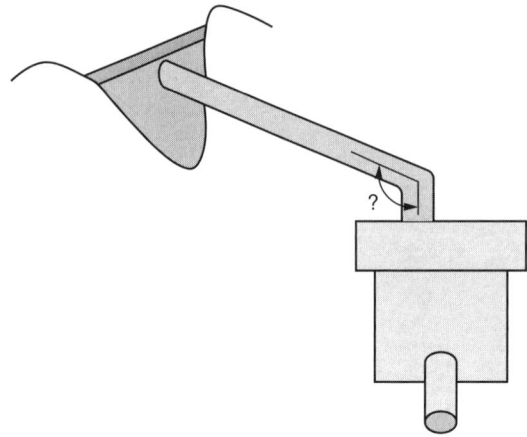

UNIT 25

Units of Length Measure

Basic Principles of Units of Length Measure

- Review the Introduction to Measurement in the front of the book.
- Review denominate numbers in Section I of the Appendix.
- Review equivalents in section II of the Appendix.
- Study the tables of units of length measure.

Many problems involve units as well as numbers. When solving many of these problems, the units must all be the same. Often, the units given in the problem are not the same, so some of the units must be converted to other units in order to solve the problem. If a unit is changed, the number with that unit will also change. The two quantities must be kept equivalent. What follows is one method to convert from one unit to another. While there are other methods, this method will *always* convert units correctly.

EXAMPLE 1:   We need to convert 36 inches to feet.

Almost everyone knows that there are 12 inches in 1 foot, so 12 plays a key part in converting between units. 12 is known as a **conversion factor**. The question that must be answered next is: Do we multiply or divide by 12? The following method will show you the correct operation to perform each time.

First, write the quantity (including units) that is to be converted

    36 inches

Next to this, write a fraction. The fraction contains the units that you are trying to get rid of, so place them in the fraction so that they can be canceled. Remember that when a number such as 36 inches is written, it is really written in the numerator of a fraction with just 1 as the denominator (which we do not write). Since the inches in 36 inches is in the numerator, our conversion fraction will have inches in the denominator so they can cancel. The numerator of the conversion fraction contains a unit that has a well-known relation to the unit you are canceling. It may not be the unit that you are trying to convert to. That may involve more than one conversion fraction. The relation between the units in the fraction should be one that you know. If that relation takes you to the unit you are seeking, great! But it does not have to be that way each time. For our problem

$$36 \text{ inches} \left(\frac{\text{feet}}{\text{inches}} \right)$$

The inches will cancel and leave feet in the numerator. The units are now taken care of. Now go to the conversion fraction and fill in the numbers dealing with the relationship in the fraction. CONCERN YOURSELF WITH ONLY WHAT IS IN THE FRACTION. (In this case feet and inches.) We know that 1 foot = 12 inches, so that is what is put into the fraction.

$$36 \text{ inches} \left(\frac{1 \text{ foot}}{12 \text{ inches}} \right)$$

This is solved as a fraction problem and we see that, in this case, we divide by 12.

$$36 \text{ inches} \left(\frac{1 \text{ foot}}{12 \text{ inches}} \right) = 3 \text{ feet}$$

Remember:

- Set up each fraction to cancel out units.
- Put numbers in the fractions to make each fraction an equivalent.
- It may take more than one conversion fraction.

Practical Problems:

- Apply the principles of units of length measure to the problems in this unit.

For problems 1–4, express in inches.

1. 2 feet   _____

2. 9 feet   _____

3. $6\dfrac{6}{4}$ feet   _____

4. $12\dfrac{1}{2}$ feet   _____

For problems 5–8, express in centimeters.

5. 3 meters   _____

6. 15 meters   _____

7. 0.4 meters   _____

8. 7.35 meters   _____

For problems 9–12, express in feet.

9. 48 inches   _____

10. 9 inches   _____

11. 35 inches   _____

12. 732 inches   _____

For problems 13–16, express in meters.

13. 3,500 centimeters   _____

14. 266 centimeters   _____

15. 8 centimeters   _____

16. 0.5 centimeters   _____

17. Express 4 feet 7 inches in inches.   _____

18. Express 8 feet 8 inches in feet.   _____

19. Express 6 meters 58 centimeters in centimeters.   _____

20. Express 2 meters 94 centimeters in meters.   _____

21.   The drop in height for water in the penstock is given in inches. What is this drop in feet?

_____

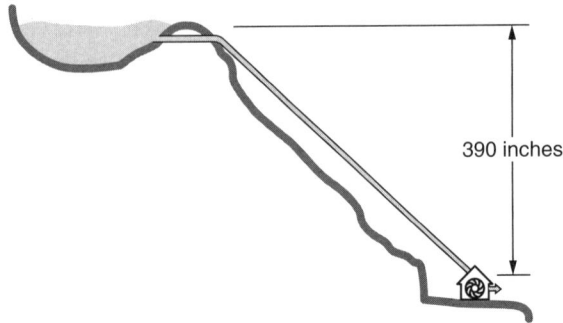

390 inches

22.   The ground clearance of a wind turbine blade is 12,000 centimeters. Express this clearance in meters.

_____

23.   The depth of a geothermal well is 275 feet 10 inches. Express the depth in feet.

_____

24.   Express the dimensions of the PV panel shown in centimeters.

_____

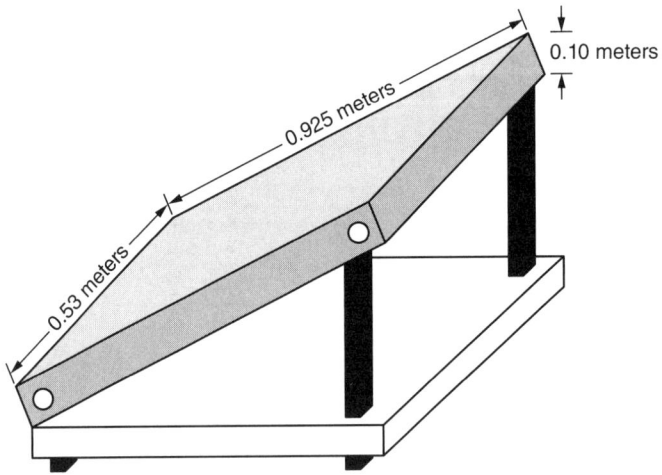

0.10 meters

0.925 meters

0.53 meters

25.   The height of a wind turbine tower is 33 feet. Express this height in inches.

_____

UNIT 26

Equivalent Units of Length Measure

Basic Principles of Equivalent Units of Length Measure

* Review the Introduction to Measurement in the front of the book.
* Review denominate numbers in Section I of the Appendix.
* Review equivalents in Section II of the Appendix.
* Study the tables of units of length measure.

In Unit 25, conversions were made between units in the US customary system. Conversions were also made between units in the metric system. The United States is one of a very, very few countries that does not use the metric system of measurement. Global trading has resulted in many manufactured goods coming into the United States that have dimensions and/or instructions relating to the metric system. There will be times when it will be desirable to convert measurements made in one unit system to the other unit system.

The method described and used in Unit 25 will also work when converting between unit systems. It may take more than one fraction to make the conversion, but the process is the same. One of the fractions will be to determine equivalents between the two unit systems.

EXAMPLE 1:   Convert 36 inches to meters.

First write the quantity (including units) that is to be converted.

> 36 inches

Before writing any fraction, take time to look at the tables listing equivalents. Look to see if there is an equivalence given between the two units that we are trying to convert between. The table in Appendix II shows a relationship between inches and meters.

145

As we had done before, write a fraction. This time the fraction contains the units and equivalent numbers that we found in the table. For our problem:

$$36 \text{ inches} \left(\frac{0.0254 \text{ meters}}{1 \text{ inch}} \right)$$

The inches will cancel and leave meters in the numerator.

$$36 \text{ inches} \left(\frac{0.0254 \text{ meters}}{1 \text{ inch}} \right) = 0.9144 \text{ meters}$$

Remember:

- Set up each fraction to cancel out units.
- Put numbers in the fractions to make each fraction an equivalent.
- It may take more than one conversion fraction.
- One fraction will have an equivalence between the two unit systems.

Practical Problems:

- Apply the principles of equivalent units of length measure to the problems in this unit.

For problems 1–6, round answers to the nearer hundredth when needed:

1. 8 inches to centimeters   _____

2. 4 feet to centimeters   _____

3. 3 feet 5 inches to centimeters   _____

4. 10 inches to meters   _____

5. 5 feet to meters   _____

6. 9 feet 7 inches to meters   _____

For problems 7–12, round answers to the nearer thousandth when needed:

7. 15.7 centimeters to inches   _____

8. 3.4 meters to inches   _____

9. 213 centimeters to feet   _____

10. 7.1 meters to feet   _____

11. 91 centimeters to feet and inches   _____

12. 4.8 meters to feet and inches   _____

Round answers to the nearer hundredth where needed.

13. A wind turbine blade measures 61.5 meters in length. What is the
    length of the blade in feet?   _____

14. What is the length of the penstock shown in meters?   _____

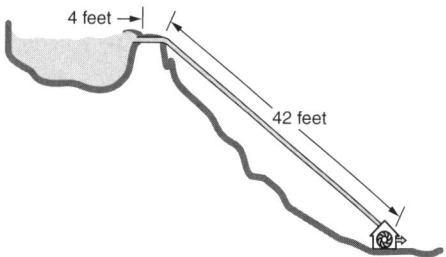

15. A solar collector array measures 24 feet 9 inches in width. What is
    the width in centimeters?   _____

16. The diameter of a wind turbine tower is 5 meters. What is this diameter
    in feet?   _____

17. What are the dimensions of the horizontal geothermal bed shown in
    centimeters?   _____

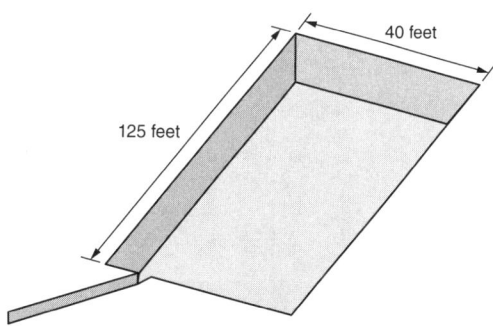

    A. Length   _____
    B. Width   _____

18. A low-head hydro generator is positioned 24.6 feet below the water surface. What is that height in meters?   _____

19. The length of a conduit down the side of a wind turbine tower is 19.5 meters. Convert this to feet.   _____

20. The length of tubing for a horizontal bed geothermal heat exchanger is 1,375 feet. Convert this to meters.   _____

21. Express the distance in meters from the ground to the cross arm shown.   _____

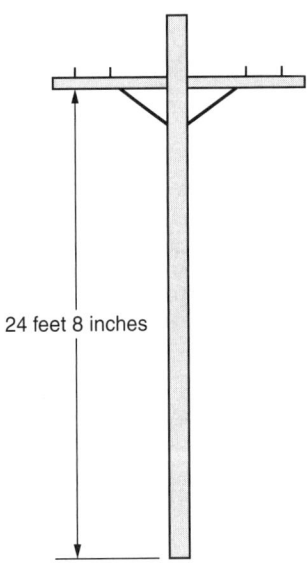

24 feet 8 inches

22. The distance between two wind generators is 240 meters. How far is this in feet and inches?   _____

23. Express the diameter of the penstock piping shown in inches.   _____

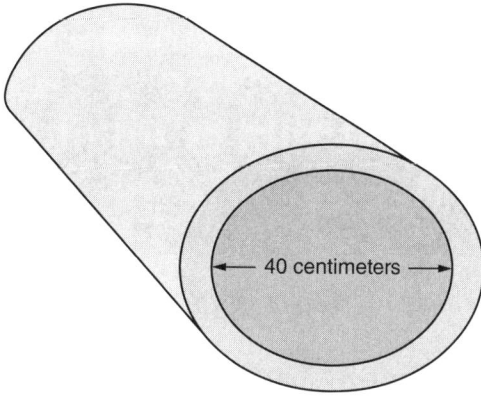

40 centimeters

24. A cable connecting a hydro generator with the main house is 62 feet
    7 inches long. What is that length in meters?   _____

25. Clearance between a solar collector and the house roof is 4 centimeters.
    What is that clearance in inches?   _____

UNIT 27

Equivalent Units of Additional Direct Measure

Basic Principles of Equivalent Units of Additional Direct Measure

- Review the Introduction to Measurement in the front of the book.
- Review denominate numbers in Section I of the Appendix.
- Review equivalents in Section II of the Appendix.

There are many units that have equivalent units in other unit systems. The trick to solving these problems is to find the set of equivalences or the combination of equivalences that will get from one unit to the other. The method is the same as we have been using in previous units. Remember that the fractions that you set up must be set up so that the unit you are converting from will get canceled.

EXAMPLE 1: Convert 2 miles per hour to meters per second.

The units miles per hour should be written as a fraction $\frac{\text{miles}}{\text{hour}}$. This is going to require multiple conversions. Take them one at a time. Start with miles. There is no conversion factor for miles to meters, but there is one converting from miles to kilometers. There is another conversion from kilometers to meters. So we now have

$$\frac{2 \text{ miles}}{\text{hour}} \left(\frac{\text{kilometers}}{\text{miles}} \right) \left(\frac{\text{meters}}{\text{kilometers}} \right)$$

This converts the miles to meters, but the units now are meters per hour, and we want meters per second. Notice that the hour is in the denominator, not in the numerator. The fractions we write to convert this part of the units need to be written to cancel the hour. Continuing now with the hour conversion.

150

$$\frac{2 \text{ miles}}{\text{hour}} \left(\frac{\text{kilometers}}{\text{miles}}\right) \left(\frac{\text{meters}}{\text{kilometers}}\right) \left(\frac{\text{hour}}{\text{minutes}}\right) \left(\frac{\text{minute}}{\text{seconds}}\right)$$

Notice that the fractions with hours are not next to each other. That is okay! Now that the units are taken care of, fill in each fraction with equivalents.

$$\frac{2 \text{ miles}}{\text{hour}} \left(\frac{1.609 \text{ kilometers}}{1 \text{ miles}}\right) \left(\frac{1,000 \text{ meters}}{1 \text{ kilometer}}\right) \left(\frac{1 \text{ hour}}{60 \text{ minutes}}\right) \left(\frac{1 \text{ minute}}{60 \text{ seconds}}\right)$$

$$= 0.89 \frac{\text{meters}}{\text{second}} \text{ (Rounded off to nearer hundredth.)}$$

Remember:

- Set up each fraction to cancel out units.
- Put numbers in the fractions to make each fraction an equivalent.
- It may take more than one conversion fraction.
- One fraction will have an equivalence between the two unit systems if needed.

Practical Problems:

- Apply the principles of equivalent units of additional direct measure to the problems in this unit.

Round answers to the nearer hundredth when needed.

1. 2 gallons to liters   _____

2. 7.25 gallons to liters   _____

3. 14 liters to gallons   _____

4. 250 liters to gallons   _____

5. 5 miles to feet   _____

6. 10.3 miles to feet   _____

7. 8,432 feet to miles   _____

8. 27,425 feet to miles   _____

9. 12 miles to kilometers   _____

10. 3.2 miles to kilometers   _____

11. 40 gallons per second to liters per second   _____

12. 350 gallons per hour to liters per hour   _____

13. 2,400 liters per minute to gallons per second   _____

14. 6,476 gallons per hour to liters per minute   _____

15. 75 feet per minute to meters per second   _____

16. A wind turbine is designed so that it does not start turning if the wind
    speed is slower than 10 kilometers per hour. Express the speed in miles
    per hour.   _____

17. What is the flow in liters per second for the solar collector shown?   _____

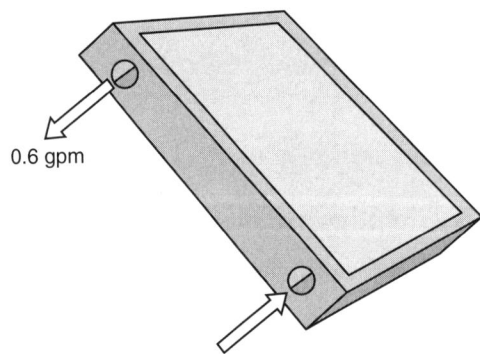

0.6 gpm

18. A stream has a flow of 250,000 gallons per hour. What is the stream's
    flow measured in liters per hour?   _____

19. What value in gallons per hour is equivalent to 5,400 gallons
    per second?   _____

20. A wind generator is rated for a wind speed of 25 kilometers per hour.
    Express this speed in miles per hour.   _____

21. The weight of an operating penstock (water and piping together) is 18 pounds per foot. Convert this to kilograms per meter. _____

22. A low-head hydro generator is rated for 130 liters per second. What is this flow in gallons per second? _____

23. A wind generator is designed to stop turning when the air speed exceeds 75 miles per hour. Express this in kilometers per hour. _____

24. A solar collection panel has a recommended coolant flow through it of 0.027 liters per second. What is this flow in gallons per minute? _____

25. A stream has a measured flow of 135,000 gallons per hour. Will this meet the minimum flow of 140 liters per second? _____

Computed Measure

UNIT 28

Computed Length Measure

Basic Principles of Computed Length Measure

- Review the Introduction to Measurement in the front of the book.
- Review denominate numbers in Section I of the Appendix.
- Review the tables of length measure in Section II of the Appendix.
- Use formulas for perimeters found in Section IV of the Appendix.

In the previous section measurements were given and you had to convert them to another unit system. There are situations where measurements have to be manipulated—added, subtracted, multiplied, and so on. Usually, when this happens, the measurement has a single unit. An exception to this can occur when adding or subtracting lengths in feet and inches. There are two ways to solve this type of problem. The first is to convert the units to all feet or all inches and then work the problem. The answer would then be converted back to a combination of feet and inches. The second way is to work with the units separately first. For example, if adding measurements that are combinations of feet and inches, add all of the inches together. If the inch total is greater than 12, convert the value to a combination of feet and inches. Add the feet number in with the other feet quantities to get the total feet value.

EXAMPLE 1:   Add 2 feet 8 inches and 3 feet 6 inches.

$$\begin{array}{r} 2 \text{ feet } 8 \text{ inches} \\ + \ 3 \text{ feet } 6 \text{ inches} \\ \hline \end{array}$$

First, the 8 inches and the 6 inches are added: 8 inches + 6 inches = 14 inches. This is greater than 12 inches, so this part can be converted to 1 foot 2 inches. Next, add that 1 foot (from the 14 inches) + 2 feet + 3 feet to give a total of 6 feet.

156

2 feet 8 inches
+ 3 feet 6 inches
5 feet 14 inches or 6 feet 2 inches

When subtracting, it may be necessary to borrow 1 foot and convert it to 12 inches before subtracting.

EXAMPLE 2: Subtract 3 feet 4 inches from 5 feet.

5 feet
− 3 feet 4 inches

4 inches needs to be subtracted from something, so borrow 1 foot from the 5 feet and convert it to 12 inches.

5 feet   4 feet 12 inches
− 3 feet 4 inches   =   − 3 feet   4 inches
                          1 foot   8 inches

Most of the time, the dimensions will need to be in one unit before being worked. The above is an exception to that rule.

One of the situations where measurements need to be manipulated is with perimeters. A **perimeter** (*P*) is the distance or length *around* the outside of a figure.

The perimeter (*P*) of a square is:

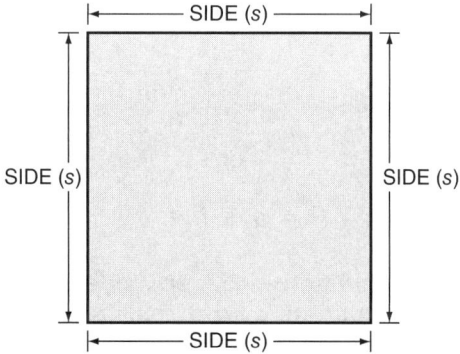

$$Perimeter = side + side + side + side$$

or

$$P = 4s$$

The perimeter of a rectangle is:

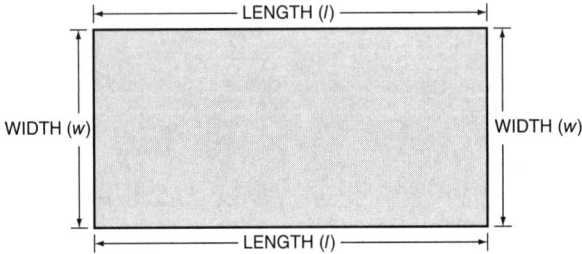

$$Perimeter = length + width + length + width$$

or

$$P = 2l + 2w$$

The perimeter of a triangle is:

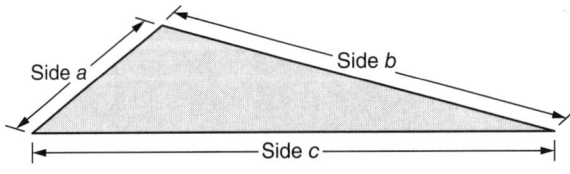

$$Perimeter = side\ a + side\ b + side\ c$$

or

$$P = a + b + c$$

The perimeter of a circle has a special name—the **circumference**. The distance from the center of the circle to the outside edge is called the **radius**. The distance from one side edge of a circle through the center to the other side edge of the circle is called the **diameter**. Diameter = 2 × radius or $D = 2r$. So:

$$Radius = \frac{diameter}{2} \text{ or } r = \frac{D}{2}$$

The circumference (*C*) of a circle is:

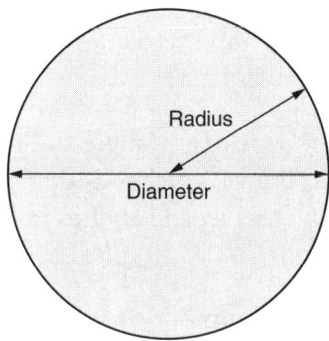

Circumference = 2 × π × radius

or

$$C = 2\pi r$$

π is a symbol used in mathematics to represent a constant value associated with circles. Unless otherwise directed, use the value 3.1416 whenever π is used.

If the diameter of a circle is given, the circumference can be found using:

$$C = \pi D$$

Practical Problems:

• Apply the principles of computed length measure to the problems in this unit.

Round answers to two decimal places, if necessary.

1. A passive skylight is being installed in the roof of a house. The skylight is a square with a side 1 foot 4 inches in length. What is the length of flashing needed around the outside of the skylight? _____

2. A solar collector is a rectangle with dimensions 8 feet 2 inches long and 3 feet 8 inches wide. A bead of caulking goes around the edge of the collector to seal the solar window to the collector. What is the length of the caulking bead needed for the collector? _____

3. What is the width of insulation that would wrap around a $1\frac{1}{2}$-inch diameter pipe coming from a solar heater?   _____

4. A wind turbine blade is 22 inches long. How far does the tip of the blade travel when it makes one revolution?   _____

5. A low-head hydro dam is in the shape of a triangle and requires fencing around the edge. The sides are 52 feet 8 inches and 43 feet 6 inches in length and the front of the dam is 35 feet 11 inches in length. What is the total length of fencing needed?   _____

6. A gasket needs to be made for a connection on the piping for the penstock for a hydro generating system. The pipe is 6 inches in diameter. What is the length of gasket around the connection?   _____

7. A wind turbine has a fence built around the base of the tower. The fence is square with a side of 952 centimeters. What is the length of the fence?   _____

8. Edging is needed around the outside of the PV panel shown. What length of edging is needed for this panel?   _____

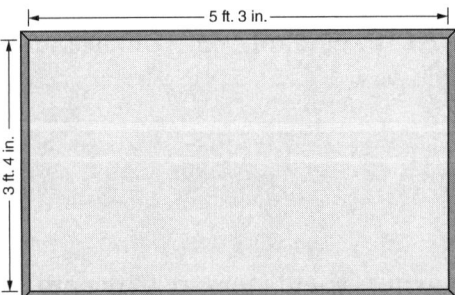

9. The hub of a wind turbine is able to rotate in a circle. How far does the tip of the hub travel when it rotates in one circle if the tip is 95 centimeters from the point of rotation?   _____

10. The trench for a geothermal heat exchanger is a U-shaped layout. The length of each of the long sides is 34 meters 74 centimeters. The length of the cross-trench is 15 meters 27 centimeters. What is the length of tubing needed for this setup?   _____

11. Determine the length of piping needed to connect the solar panel on the roof with the storage tank in the basement.   _____

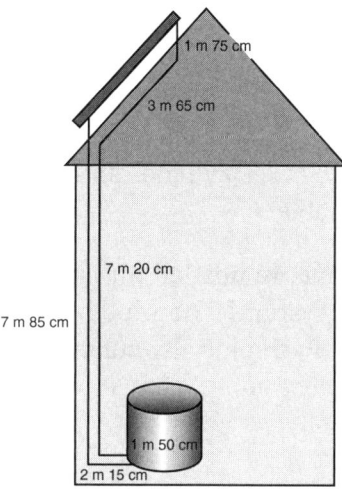

12. Felt is wrapped around the shaft of a wind turbine where it enters the casing. What length of felt is needed for a shaft that is 2 inches in diameter?   _____

13. Find the total length of piping for the penstock shown.   _____

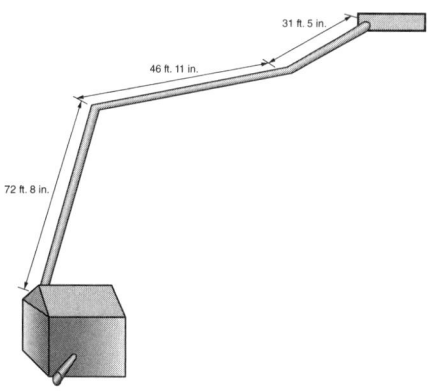

14. The casing for a wind turbine is rectangular in shape. What is the length of gasket needed to seal the upper to the lower half of the casing if the casing is 45 feet 5 inches long and 16 feet 8 inches wide?   _____

15. A road to a wind farm has segments of 2 miles 3,280 feet, 1 mile 4,100 feet, and 3,650 feet. How long is the road to the wind farm?   _____

16. Electrical cables totaling 133 feet 2 inches in length connect 3 PV solar arrays. Two of the cables are 54 feet 7 inches and 31 feet 9 inches in length. How long is the third cable?   _____

17. Two different lengths of piping are needed for an open system for a geothermal heating/cooling system. The pipe lengths are 134 feet 9 inches and 172 feet 2 inches. Each pipe also runs a distance of 23 feet 4 inches from the well to the heat pump. How much piping is needed for this system?   _____

18. Owner Nell wants to put a fence around her wind farm and post danger notices. The area to be fenced is a square with a side 654 feet 3 inches on a side. Find the length of fence needed.   _____

19. Technician Carson had a piece of conduit 14 feet long on his truck. He had to cut a piece of conduit 6 feet 7 inches long for a repair job. How much conduit does he have left?   _____

20. Manager Suzanne has 2 hydro systems that connect to a distribution system. The 2 generators are 4 meters 75 centimeters and 166 meters 43 centimeters away from the distribution panel. Suzanne must run 2 wires from each generator to the panel. How much wire does Suzanne need?   _____

21. A heat exchanger bed for a geothermal heating system has circular coils in it. The coils are 9 feet in diameter. There are 17 coils in the bed. How many feet of tubing are in the bed?   _____

22. A hold-down strap is needed to hold a 6-inch diameter penstock pipe in place. How long does the strap have to be?

_____

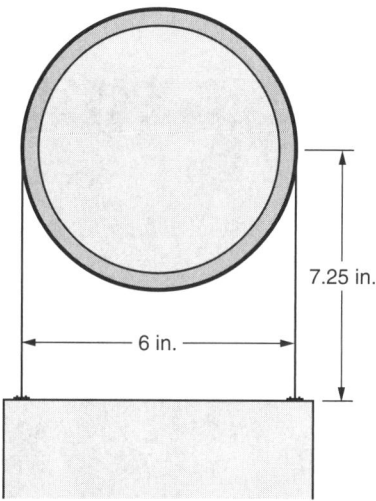

7.25 in.

6 in.

NOTE: Refer to Section III of the Appendix on how to read a vernier caliper or micrometer for problems 23–25.

23. A shaft for a generator turbine is to be exactly 2.5 inches in diameter. The diameter is measured using a caliper. How much wear has occurred on the shaft?

_____

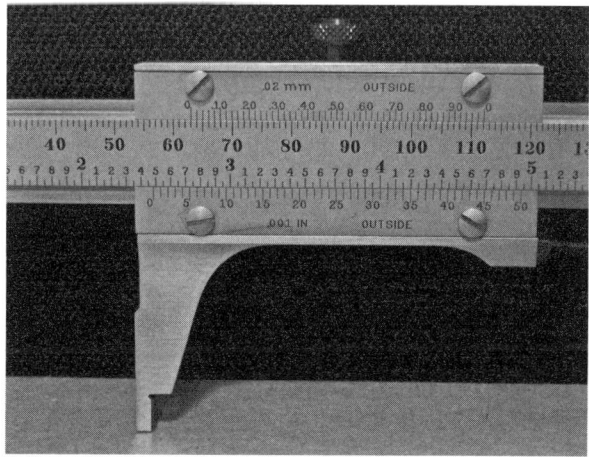

24. End play is the movement of a shaft along the axis (length) of the shaft. Allowable end play on a particular wind turbine shaft is 0.02 centimeters. The end play movement is measured using a caliper and is shown. Is this end play allowable?

_____

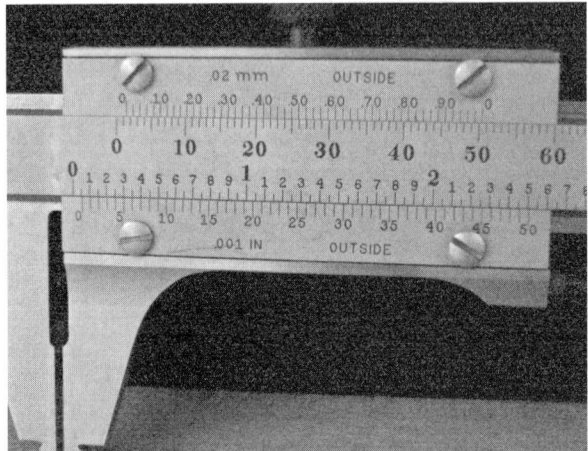

25. A second shim is needed to level a hydro generator. It must match the shim measured using a micrometer. The reading on the barrel is shown in the diagram. How thick must the shim be?

_____

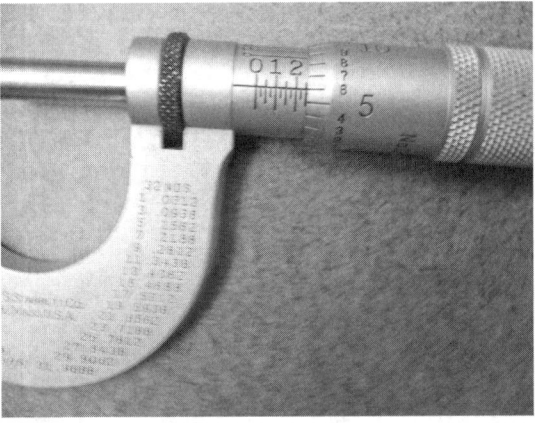

UNIT 29

Area Measure

Basic Principles of Area Measure

- Review the Introduction to Measurement in the front of the book.
- Review denominate numbers in Section I of the Appendix.
- Use formulas for areas found in Section IV of the Appendix.

The amount of space on the surface of a figure is called the *area*. Area is also the number of square units equal in measure to the surface of a figure. A number of figures have regular shapes to them. Formulas have been developed that can be used to find the area of these regular figures. The various formulas given below and in Section IV of the Appendix can be used throughout this unit.

To solve for an area (*A*), determine the correct formula to use based on the shape of the figure. Substitute values for the terms in the formula and solve for the answer. The units for area are square units—square inches, square feet, square centimeters, square meters, and so on. The way to obtain square inches is by multiplying inches by inches. The way to get square feet is by multiplying feet by feet. Therefore, care must be taken to be sure that the dimensions of the figure whose area is being determined have the same units. Do not multiply a dimension in feet by one in inches and so on.

Helpful Hint: When solving any of these problems, make a little table of what dimension you have and what its measurement is. Then, write the correct formula, substitute into it using the values in your table, make sure all of the units are the correct ones needed, and then solve the problem.

The areas for different regular figures can be found using the following formulas.

The area of a square is:

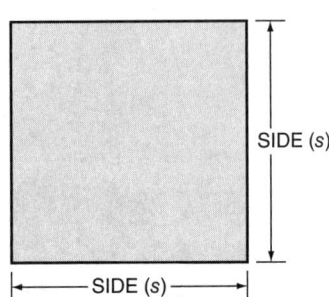

$$Area = side \times side$$

or

$$A = s \times s$$

$s \times s$ can also be written as s^2 (s squared).

So $A = s \times s$

$$= s^2$$

The area of a rectangle is:

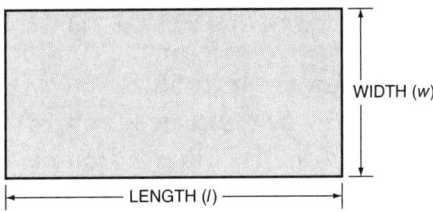

$$Area = length \times width$$

or

$$A = l \times w$$

NOTE: The length or the width of a rectangle can be found if the area and one of the dimensions are known. The above formula can be rearranged to find the unknown value.

$$l = \frac{A}{w} \text{ or } w = \frac{A}{l}$$

The area of a circle is:

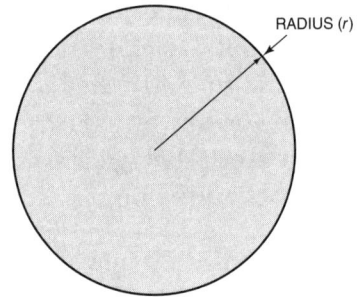

$$Area = \pi \times radius \times radius$$

or

$$A = \pi \times r \times r$$

$$= \pi \times r^2$$

As in the last unit, use $\pi = 3.1416$.

$$A = 3.1416 \times r^2$$

NOTE: The radius must be used here. Radius × radius is **NOT** the diameter. If the diameter is given, divide it by 2 to get the radius and then use that value for the formula.

The area of a triangle is:

$$\text{Area} = \frac{1}{2} \times \text{base} \times \text{height}$$

or

$$A = \frac{1}{2} \times b \times h$$

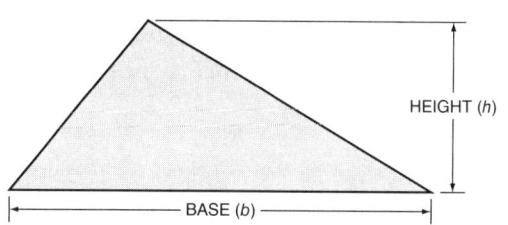

HEIGHT (h)

BASE (b)

NOTE: The base b and height h must form an angle of 90°. (They must be perpendicular to each other. One of the dimensions does not have to be a side of the triangle - the only time it would be is for a right triangle.)

NOTE: The length of the base or the height of a triangle can be determined if the area and the other dimension are known. The area formula can be rearranged to find the unknown value.

$$b = \frac{2 \times A}{h} \text{ or } h = \frac{2 \times A}{b}$$

EXAMPLE 1:   Find the area of a circle with a radius of 2 inches.

$$r = 2 \text{ inches}$$

$$A = \pi \times r^2$$

$$= 3.1416 \times (2 \text{ inches})^2$$

$$= 3.1416 \times 4 \text{ square inches}$$

$$= 12.5664 \text{ square inches}$$

EXAMPLE 2:   A rectangle has an area of 24 square feet. Its width is 4 feet. Find the length of the rectangle.

$$A = 24 \text{ square feet} \qquad w = 4 \text{ feet}$$

$$l = \frac{A}{w}$$

$$= \frac{24 \text{ square feet}}{4 \text{ feet}}$$

$$= 6 \text{ feet}$$

Helpful Hint: There may be times when finding the area of a complex figure can be done by breaking the figure into two or more simple figures and working with the areas by adding or subtracting them.

EXAMPLE 3: A window on the side of a house to be used as part of a solar collector system is shown. Find the area of the window.

The area can be divided into two figures, a rectangle and a triangle.

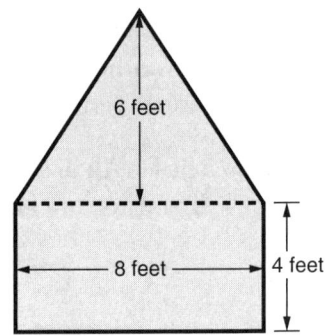

For the area of the rectangle:

$$l = 8 \text{ feet} \qquad w = 4 \text{ feet}$$

$$A = l \times w$$

$$= 8 \text{ feet} \times 4 \text{ feet}$$

$$= 32 \text{ square feet}$$

For the area of the triangle:

$$b = 8 \text{ feet} \qquad h = 6 \text{ feet}$$

$$A = \frac{1}{2} \times b \times h$$

$$= \frac{1}{2} \times 8 \text{ feet} \times 6 \text{ feet}$$

$$= 4 \text{ feet} \times 6 \text{ feet}$$

$$= 24 \text{ square feet}$$

$$\text{Total area} = 32 \text{ square feet} + 24 \text{ square feet}$$

$$= 56 \text{ square feet}$$

Practical Problems:

• Apply the principles of area measure to the problems in this unit.

1. A geothermal horizontal heat exchange bed is in the shape of a rectangle. The $2\frac{1}{2}$-ton system requires a bed that is 28 feet wide and 115 feet long. What surface area does the heat exchange bed cover?   _____

2. An upright, cylindrical, solar hot water system storage tank is 24 inches in diameter. What floor area does this tank occupy?   _____

3. To add a solar skylight to a home, a square hole 14 inches on a side is cut into the roof. What roof area does this skylight occupy?   _____

4. The dam for a low-head hydro system has backed water up, forming a triangular-shaped lake. What is the surface area of the lake formed?   _____

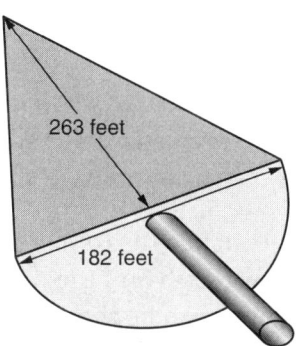

263 feet

182 feet

5. A wind turbine has blades that are 1.8 meters long. What is the area of the circle swept out by the rotor blades?   _____

6. A rectangular solar collector measures 1.25 meters wide and 2.8 meters long. What is the surface area of this solar collector?   _____

7. A wind turbine has a fence built around the base of the tower. The fence forms a square with a side of 12.25 meters. What is the area enclosed by the fence?   _____

8. A penstock is composed of 5-inch diameter pipe. What is the cross-sectional area of the penstock? _____

9. A roof measures 46 feet 6 inches long and 16 feet 4 inches wide. What is the maximum area for potential solar PV panels? _____

10. A geothermal heat exchange bed is square in shape, with the length of a side being 52 feet 6 inches. What is the surface area of the bed? _____

11. A large wind turbine has a rectangular housing measuring 42 feet long and 14 feet wide. What is the surface area for the top of the turbine? _____

12. The discharge line from a hydro turbine is a pipe 8 inches in diameter. What is the cross-sectional area of the discharge pipe? _____

13. A PV panel is a square 2 meters 15 centimeters on a side. What is the collector area? _____

14. How much area must be fenced off if a circular fence of radius 8 feet 2 inches is put around a wind turbine tower? _____

15. A road built to a wind farm is to be surfaced. The road is 0.8 miles long and will be 16 feet wide. What is the surface area of the road? _____

16. The support tower for a wind turbine needs to be repainted. What is its surface area? The tower is 45 feet high and the diameter of the circular tower is 4 feet. **Hint:** The shape can be thought of as a rectangle. The length of the side is the circumference of the circle. _____

17. The blades of a low-head hydro turbine are triangular in shape. Each blade is 3 inches from the axle to the outside edge, and the outside edge is 2.75 inches long. There are 8 blades on the turbine. What is the total area of the turbine blades? _____

18. A college has installed a large PV array on a field on its campus. The array consists of rectangular panels 5 feet 6 inches long and 3 feet 4 inches wide. The array is 15 panels long and 12 panels wide. What area does this array cover? _____

19. Air causing a wind turbine to turn is disturbed as it passes the blades of the turbine. So that this does not change the airflow at the next

turbine, an area around each turbine is designated where no other turbine will be built. The no-build area is to be 6 blade lengths from the turbine. Since the turbine can spin in any direction, a circle centered on the turbine tower is designated as a no-build zone. What area on the ground would this cover for a 9-meter blade length?

20. An exit channel for water leaving a low-head hydro generator is rectangular in shape. The cross-sectional area for the channel is 206.25 square centimeters. The channel is 12.5 centimeters high. How wide is the channel?

21. Owner Nell has built a 12,400 square foot warehouse to store parts for her wind farm. The building is 80 feet wide. How deep is the warehouse?

22. The wall of a house has the shape shown and is to be fitted with windows to enclose a solar collector. What is the area of the glass needed?

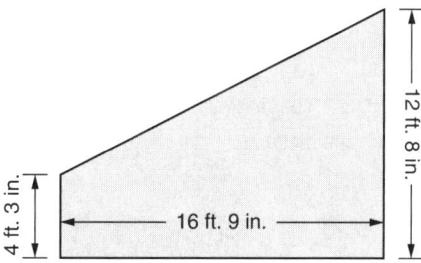

23. A house roof is used for a PV system. How much area is not covered by PV panels?

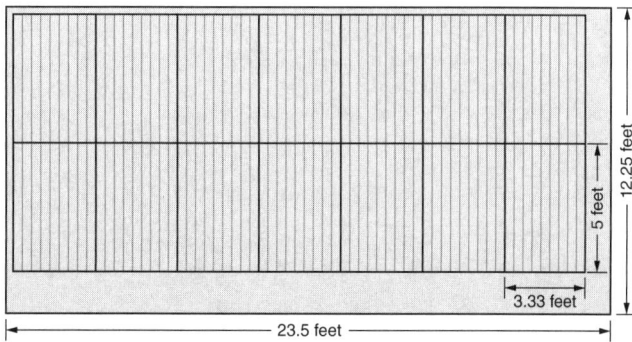

24. A solar hot water storage tank is to be insulated. What is the surface area of the top and side to be insulated? _____

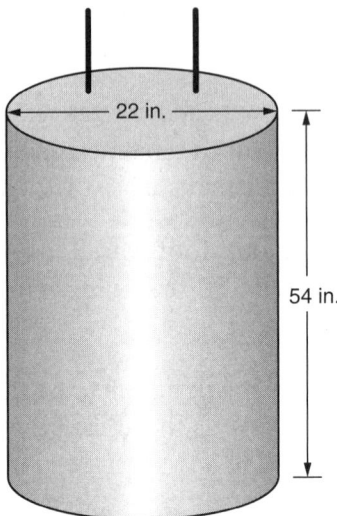

22 in.

54 in.

25. A low-head hydro dam breast is constructed as a half circle. What is the surface area of the dam breast shown? _____

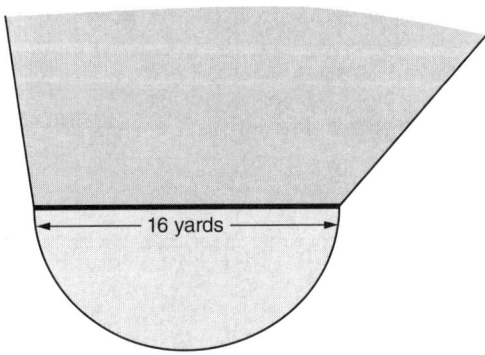

16 yards

UNIT 30

Basic Principles of Volume Measure

* Review denominate numbers in Section I of the Appendix.
* Use formulas for volumes found in Section IV of the Appendix.

The amount of space enclosed by a three-dimensional figure is called the ***volume***. Volume is also the number of cubic units equal in measure to the total space occupied by a figure. A number of figures have regular shapes to them. Formulas have been developed that can be used to find the volume of these regular figures. The various formulas given below and in Section IV of the Appendix can be used throughout this unit.

To solve for a volume (V), determine the correct formula to use based on the shape of the figure. Substitute values for the terms in the formula and solve for the answer. The units for volume are cubic units—cubic inches, cubic feet, cubic centimeters, cubic meters, and so on. The way to obtain cubic inches is by multiplying inches by inches by inches. Another way to obtain cubic inches is by multiplying square inches by inches. The way to get cubic feet is by multiplying feet by feet by feet or by multiplying square feet by feet. Therefore, care must be taken to be sure that the dimensions of the figure whose area is being determined have the same units. Do not multiply a dimension in feet by one in inches and so on.

In the last unit it was stated that to find the area of a triangle, the base and the height had to be at right angles to each other. It is a similar situation for volumes. The three dimensions **MUST** be at right angles to each other. Care should be taken to make sure the dimensions are at right angles to each other before solving the problem.

Helpful Hint: When solving any of these problems, make a little table of what dimensions you have and what their measurements are. Then, write the correct formula, substitute into it using the values in your table, make sure all of the units are the correct ones needed, and then solve the problem.

The volumes for different regular figures can be found using the following formulas.

The volume of a cube is:

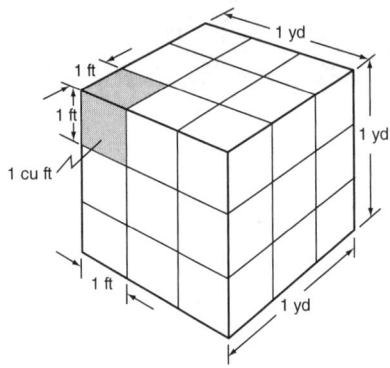

Volume = side × side × side

or

$V = s \times s \times s$

$s \times s \times s$ can also be written as s^3 (s cubed).

So $V = s \times s \times s$

$\quad = s^3$

The volume of a rectangular solid is:

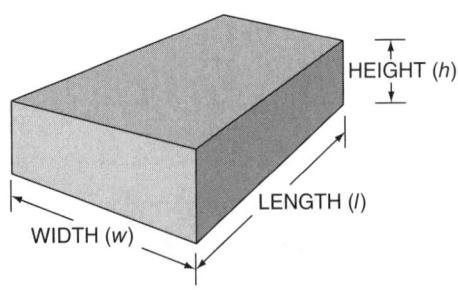

Volume = length × width × height

or

$V = l \times w \times h$

NOTE: The length or the width or the height of a rectangular solid can be found if the volume and two of the dimensions are known. The above formula can be rearranged to find the unknown value.

$l = \dfrac{V}{w \times h}$ or

$w = \dfrac{V}{l \times h}$ or $h = \dfrac{V}{w \times l}$

A right circular cylinder is a volume like a soup can. The top is a circle and height is perpendicular (at right angles) to that circle.

The volume of a right circular cylinder is:

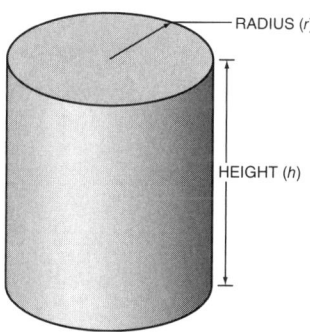

Volume = Area of a circle × height

Volume = π × radius × radius × height

or

$V = \pi \times r \times r \times h$

$= \pi \times r^2 h$

As in the last unit, use π = 3.1416.

$$V = 3.1416 \times r^2h$$

NOTE: The radius must be used here. Radius × radius is **NOT** the diameter. If the diameter is given, divide it by 2 to get the radius and then use that value for the formula.

The volume of a right triangular solid is:

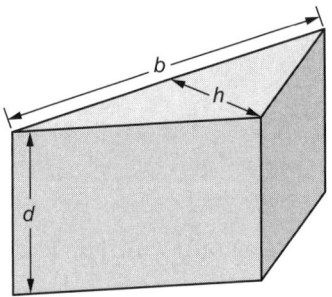

$$\text{Volume} = \frac{1}{2} \times \text{base} \times \text{height} \times \text{depth (or length)}$$

or

$$V = \frac{1}{2} \times b \times h \times d \text{ (or } l\text{)}$$

NOTE: The base b and height h must form an angle of 90° and the depth (or length) must be perpendicular to both b and h. (They must all be perpendicular to each other.)

NOTE: The length of the base or the height of a triangle or the depth of the solid can be determined if the volume and two other dimensions are known. The volume formula can be rearranged to find the unknown value.

$$b = \frac{2 \times V}{h \times d} \text{ or } h = \frac{2 \times V}{b \times d} \text{ or } d = \frac{2 \times V}{b \times h}$$

EXAMPLE 1:   Find the volume of a rectangular solid 2 feet long, 5 inches wide, and 9 inches high.

$l = 2$ feet $w = 5$ inches $h = 9$ inches

The volume of a rectangular solid is found using the formula

$$V = l \times w \times h$$

The dimensions should all have the same units, so first convert 2 feet to inches.

$$2 \text{ feet} \times \left(\frac{12 \text{ inches}}{1 \text{ foot}} \right) = 24 \text{ inches}$$

So, l = 24 inches w = 5 inches h = 9 inches

V = 24 inches $\times$ 5 inches $\times$ 9 inches

Estimating:

$$V = 20 \times 5 \times 10$$

$$= 1{,}000$$

Doing the actual multiplication:

V = 24 inches $\times$ 5 inches $\times$ 9 inches

= 1,080 cubic inches   (inches $\times$ inches $\times$ inches = cubic inches)

EXAMPLE 2: A tank is the shape of a right circular cylinder 3 feet in diameter and 5 feet high. What is the volume of the tank?

Diameter = 3 feet h = 5 feet

The volume for a right circular cylinder is found using:

$$V = \pi \times r^2 h$$

Radius: $r = \dfrac{1}{2} \times \text{diameter} = \dfrac{1}{2} \times 3 \text{ feet} = \dfrac{3}{2} \text{ feet or } 1.5 \text{ feet}$

NOTE: Since π is a decimal quantity, it is easier to use all decimals in this problem rather than decimals and fractions.

$$V = \pi \times r^2 h$$

$$= 3.1416 \times 1.5 \text{ feet} \times 1.5 \text{ feet} \times 5 \text{ feet}$$

Estimating the answer:

$$V = 3 \times 2 \times 2 \times 5$$

$$= 60$$

Solving the problem:

$$V = 3.1416 \times 1.5 \text{ feet} \times 1.5 \text{ feet} \times 5 \text{ feet}$$

$$= 35.343 \text{ cubic feet}$$

Helpful Hint: There may be times when finding the volume of a complex figure can be done by breaking the figure into two or more simple figures and working with the volumes by adding or subtracting them.

Practical Problems:

• Apply the principles of volume measure to the problems in this unit.

1. A rectangular geothermal horizontal heat exchange bed has been excavated. The bed is 42 feet wide and 92 feet long. The bed is 7 feet deep. What volume of dirt has been removed to make the bed?   _____

2. A geothermal heat exchange well has been dug. The well is 6 inches in diameter and 940 feet deep. What is the volume of dirt removed in making the well?   _____

3. A pallet of bags of geothermal heat exchange grout is a cube with a side 3 feet 6 inches long. What volume does the pallet occupy?   _____

4. The dam breast for a low-head hydro system has an earthen wall as shown. What is the volume of dirt in the earthen wall?   _____

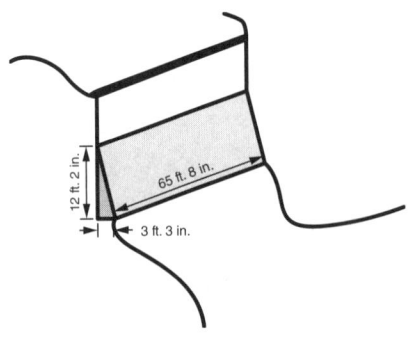

5. A cubic concrete pad is poured to support a transformer to connect a wind farm with the electrical grid. A side to the cube is 2.75 meters. What is the volume of concrete needed?

6. What is the volume of the solar collector shown?

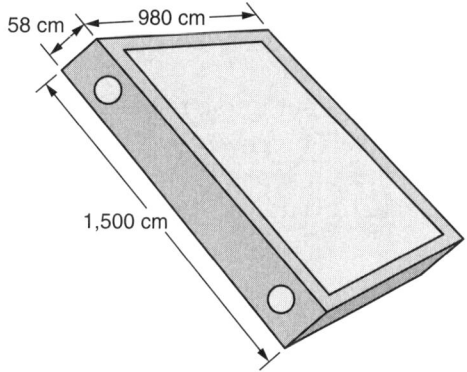

7. The penstock for a low-head hydro system is 45 feet long. The pipe is 6 inches in diameter. What volume of water would fill the pipe?

8. The bed of a pickup truck measures 99 inches long, 72 inches wide, and 23 inches deep. It is needed to carry stone for a geothermal well. What volume of stone will fit into the bed of the truck?

9. What is the volume of a 12.5-centimeter diameter well drilled 175 meters deep?

10. A low-head hydro dam is in the shape of a triangle as shown. What is the volume of water held in the dam?

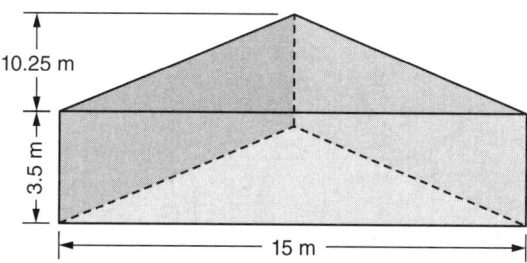

11. A geothermal heat exchange bed is a square 75 feet on a side. The bed is 12 feet deep. What volume of dirt had to be removed for this bed? _____

12. The air passing a wind turbine can be thought of as a circular cylinder of air. The area is the circle made by the turbine blade and the length is the distance the air travels in a given amount of time. On a certain day, wind travels 25 feet in one second. How much air will pass a wind turbine with 17-foot-long blades each second? _____

13. What volume of stone is needed to put into a rectangular geothermal bed 24 feet by 32 feet to a depth of 14 inches? _____

14. What is the volume of a turbine tower base that is 18 feet in diameter and 12 feet deep? _____

15. A building is being constructed to house a low-head hydro generator. The building is 12 feet by 14 feet by 8 feet. What is the volume of this building? _____

16. A PV panel has dimensions of 65.5 inches by 39.2 inches by 2 inches. What volume would a pallet of 22 of these panels occupy? _____

17. The housing of a wind turbine generator is in the shape of a cube 3.25 meters on a side. What is the volume of this housing? _____

18. A geothermal heat exchange well is 8 inches in diameter and 375 feet deep. What is the volume of the well? _____

19. A square geothermal heat exchange bed is 84 feet on a side. What volume of stone is needed to fill the bed to a thickness of 18 inches? _____

20. A large wind turbine needs a rectangular concrete pad shown. What volume of concrete is needed for this pad? _____

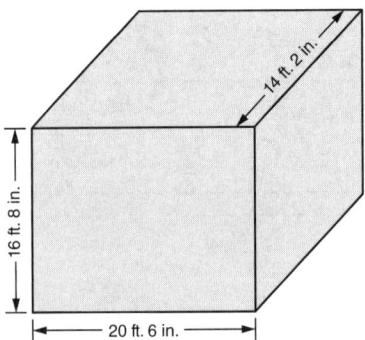

21. During the winter, a solar hot water heater must be emptied at night so the water does not freeze. A heater panel has 180 centimeters of tubing that is 5.6 centimeters in diameter. What volume of water must be drained when emptying the panel?

22. A low-head hydro generator has a 10-inch pipe for the penstock. How fast must the water flow to generate 2,500 gallons per minute?

23. The surface area of a well is 49.88 square inches after the area for the tubing has been subtracted. A batch of grout has a volume of 8,316 cubic inches. What well depth will one batch of grout fill?

24. An 8-inch diameter well is 315 feet deep. It has two $1\frac{1}{2}$-inch diameter tubes running the length of the well. The rest of the space needs to be filled with stone to help with the heat transfer. What volume must be filled with stone?

25. A penstock support is built as shown. What is the volume of concrete needed for this support? Hint: Subtract a half circle from a rectangle to get the front area and then multiply that by the thickness.

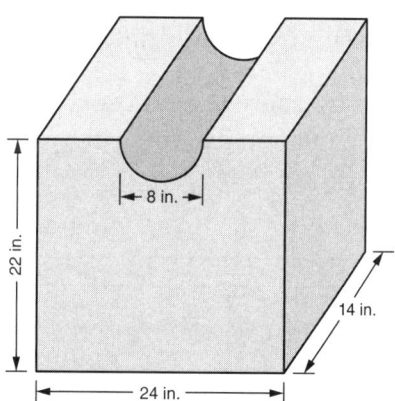

UNIT 31

Equivalent Units of Area and Volume Measure

Basic Principles of Equivalent Units of Area and Volume Measure

- Review denominate numbers in Section I of the Appendix.
- Study the tables of equivalent units of area and volume measure in this unit and Section II of the Appendix.
- Use formulas for areas and volumes found in Section IV of the Appendix.

Converting from one set of units of area or volume measure to another is done in the same manner as conversion of linear measure. Fractions are formed to remove the current unit of measure and replace it with the new unit. The only differences are the equivalent units.

There are 3 feet in 1 yard. For area measure the unit is square feet and square yards. What is the relationship between them? The relation can be seen by looking at a diagram of a square of 1 yard on a side.

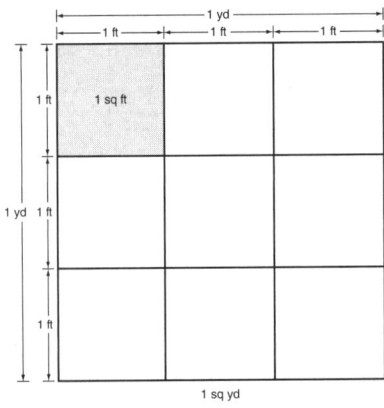

Divide each side into 1-foot lengths. Making 1-foot squares divides the area into 9 squares. So 1 square yard is equivalent to 9 square feet. That happens to be 3 feet × 3 feet = 9 square feet.

In a similar manner, a cubic yard is a cube 1 yard on a side. Dividing each side into 3 feet generates 27 cubes of size 1 cubic foot.

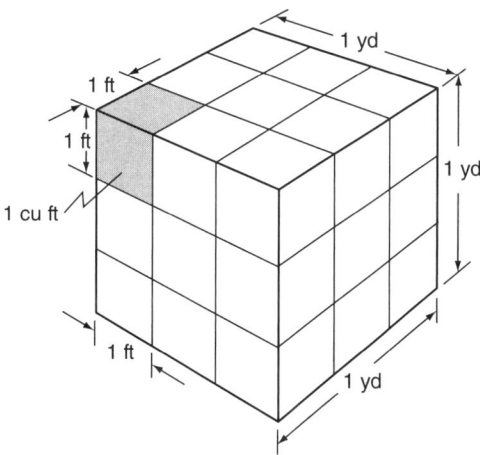

1 cubic yard equals 27 cubic feet. This is 3 feet × 3 feet × 3 feet = 27 cubic feet.

All of the area equivalences and the cubic equivalences can be determined in a similar manner. The tables below give equivalences.

ENGLISH AREA MEASURE		
1 square yard (sq yd)	=	9 square feet (sq ft)
1 square foot (sq ft)	=	144 square inches (sq in)

METRIC AREA MEASURE		
100 square millimeters (mm²)	=	1 square centimeter (cm²)
10,000 square centimeters (cm²)	=	1 square meter (m²)
1,000,000 square meters (m²)	=	1 square kilometer (km²)

1 square meter (m²)	=	10.763910 square feet (sq ft)
1 square meter (m²)	=	1,550.000 square inches (sq in)
1 square centimeter (cm²)	=	0.155000 square inch (sq in)
1 square millimeter (mm²)	=	0.001550 square inch (sq in)

1 square foot (sq ft)	=	0.092903 square meter (m²)
1 square inch (sq in)	=	0.000645 square meter (m²)
1 square inch (sq in)	=	6.451600 square centimeters (cm²)
1 square inch (sq in)	=	645.160 square millimeters (mm²)

ENGLISH VOLUME MEASURE		
1 cubic yard (cu yd)	=	27 cubic feet (cu ft)
1 cubic foot (cu ft)	=	1,728 cubic inches (cu in)

METRIC VOLUME MEASURE		
1,000 cubic millimeters (mm³)	=	1 cubic centimeter (cm³)
1,000,000 cubic centimeters (cm³)	=	1 cubic meter (m³)

ENGLISH/METRIC VOLUME EQUIVALENCES		
1 cubic foot (cu ft)	=	28,317 cubic centimeters (cm³)
1 cubic foot (cu ft)	=	0.028317 cubic meter (m³)
1 cubic meter (m³)	=	61,023 cubic inches (cu in)
1 cubic meter (m³)	=	35.314667 cubic feet (cu ft)

EXAMPLE 1:   A low-head hydro dam contains 3,676 cu ft of water. How many gallons of water does this dam hold?

From the table of English Volume Measure Equivalents in Section II of the Appendix, the relation between gallons and cubic feet is 1 gal = 0.133681 cu ft.

$$3{,}676 \text{ cu ft} \left(\frac{1 \text{ gal}}{0.133681 \text{ cu ft}} \right) = 27{,}498.3 \text{ gal (Rounded off)}$$

Practical Problems:

- Apply the principles of equivalent units of area and volume measure to the problems in this unit, rounding answers to the nearer hundredth, if needed.

1. Nell's Wind Farm is composed of 620,730 square feet. Express this area in acres.

2. A roof being considered for a solar PV system has an area of 882 square feet. What is this area in square meters?   _____

3. The cross-section of a low-head hydro turbine penstock is 28.27 square centimeters. Convert this area to square inches.   _____

4. A PV panel is shown. What is the area of the panel in square feet?   _____

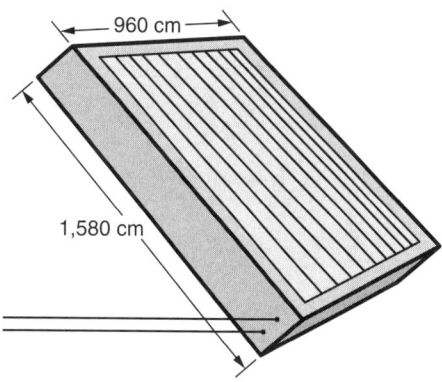

5. A low-head hydro dam has a surface area of 0.75 acres. What is the area of the dam in square feet?   _____

6. A wind turbine is to have a circle of free space around the turbine tower as shown. Determine the area in square feet.   _____

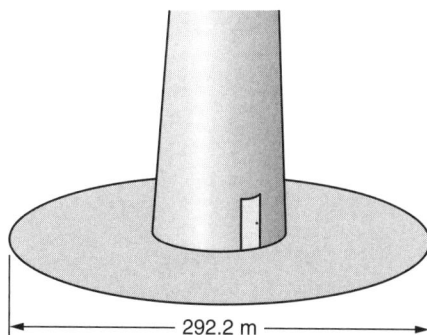

7. The sweep area (the area passed through by the blades of a wind turbine) of a turbine is 7,608.6 square feet. Convert this to square inches.   _____

8. The surface area of a solar PV panel is 2.31 square yards. Convert this to square feet. _____

9. A rectangular geothermal heat exchange bed is shown. What is the bed's area in acres? Hint: Find the area in square meters; convert first to square feet, then to acres. _____

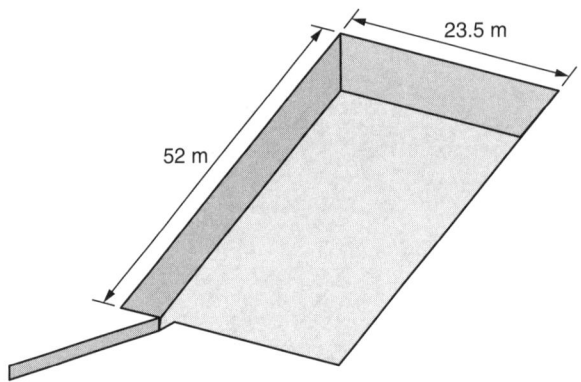

52 m

23.5 m

10. The surface of penstock piping must be painted. The surface area of the piping is 3,500 square inches. How many square meters must be covered? _____

11. A 1.5-acre lot is to be covered with solar PV panels that measure 76 inches by 40 inches. What is the largest number of panels that could be arranged on a lot that size? (The dimensions of the lot can be adjusted as long as the area remains 1.5 acres.) _____

12. A covered storage area is 2 acres in size. It is 425 feet wide. How deep is the storage area? _____

13. Solar energy reaching the ground one day is measured at 15,285 watts/ square meter. A solar heat collector is shown. How many watts of solar energy fall on the collector?

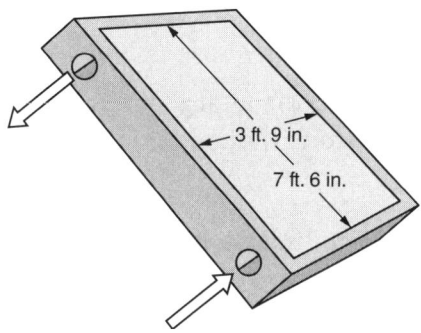

14. A geothermal well has a volume of 7.5 cubic meters. What is the volume in cubic feet?

15. Water is flowing in a low-head penstock at a rate of 52 gallons per second. What is this flow rate in cubic feet per second?

16. One batch of geothermal well heat transfer grout has a volume of 5.55 cubic feet. Convert this to cubic meters.

17. A concrete pad is poured for the base of a wind turbine. The pad used 3.5 cubic meters of concrete. The concrete is normally ordered in cubic yards. How many cubic yards are needed?

18. A horizontal geothermal heat exchange bed was created by removing 27,500 cubic yards of soil. How many cubic meters of soil were removed?

19. A pallet of PV panels has a volume of 3.9 cubic meters. What is the volume in cubic feet?

20. A 15-centimeter-diameter well is 110 meters deep. How many cubic yards of grout would be needed to fill the well?

21. 3.5 cubic yards of concrete are used to construct a base pad for a wind turbine. The pad is a square 1.75 meters on a side. How many yards deep is the pad?   _____

22. A low-head hydro dam has 25,000 gallons of water in it. How many cubic meters is that?   _____

23. A 400 feet by 350 feet rectangular horizontal geothermal heat exchange bed had 11,400 cubic yards of stone dumped into it. The stone filled the bed to what depth?   _____

24. One gallon of water weighs 8 pounds. A low-head hydro generator penstock is made of piping 0.2 meters in diameter and 250 meters long. How much does the water weigh that fills the penstock? Hint: Find the volume of the penstock in cubic feet and convert it to gallons first.   _____

25. A stream with a flow of 156,000 gallons per hour is being considered to be used to drive a low-head hydro generator. Will this stream be able to meet the minimum required flow for the generator of 170 liters per second?   _____

SECTION

Formulas

UNIT 32

Electrical Relationships I

Basic Principles of Electrical Relationships

- Use the electrical formulas found in Section IV of the Appendix.

Formulas are relationships between specific quantities. Formulas are designed to be used over and over again by changing the values that are put in for the various quantities. Formulas are composed of letters. The letters represent specific quantities. The numbers for these quantities can change, but it is always the same quantity. To solve the formula, numbers are substituted in for the letters and the equation is worked. Care must be taken to be sure that the quantities have the correct units when they are substituted into the formulas.

You have already been using and working with formulas. Formulas were used to find perimeters, areas, and volumes in the previous units. In this unit, electrical formulas will be studied.

Current is the flow of electrons moving from one point to another. It is the flow of electric charge. The unit of measure for electric current is *amperes*. The symbol used for current is I.

Current flows as a result of an energy difference. The energy difference is known as *voltage*. This is the driving force that causes current to flow. The unit for voltage is volts, and its symbol is E.

Current flow is opposed by a quantity termed *resistance*. Resistance has units of ohms. Its symbol is R. Voltage is needed to force current through resistance. In order for current to flow as a result of having a voltage, a complete path for the current must be present. The complete path is known as a *circuit*.

The relationship between voltage, current, and resistance is known as *Ohm's Law*. This can be written as:

$$\text{Voltage } (E) = \text{Current } (I) \times \text{Resistance } (R)$$

This relationship states that voltage is directly proportional to current or directly proportional to resistance. What this means is that if voltage increases (keeping resistance constant), current will increase, or if voltage increases (keeping current constant), resistance will increase. If voltage is kept constant, current is inversely proportional to resistance. That is, as current increases, resistance decreases, or if current decreases, resistance increases. The units for these quantities are different. The unit for voltage is volts, abbreviated by V. The unit for current is amperes, abbreviated by amps. The unit for resistance is ohms, abbreviated by Ω. This means that an ampere times an ohm equals a volt.

The method to use when working with formulas is to first make a table showing the quantities that you have and what quantity you are looking for. Next, determine the formula that relates the quantities. Write the formula and rearrange it if needed to wind up with the unknown quantity by itself on the left side of the equal sign. When manipulating the formula, the same operation must be done to both sides of the equal sign. Finally, substitute the numbers for the letters (make sure the units are correct) and solve the problem.

Let us look at Ohm's Law.

$$E = I \times R$$

If voltage is what is being solved for, this formula is fine. Substitute in for I and R and solve the problem. However, if I is the unknown, the formula must be rearranged.

$$E = I \times R \qquad \text{Rearrange.}$$

$$I \times R = E \qquad \text{Divide both sides by } R.$$

$$\frac{I \times R}{R} = \frac{V}{R}$$

$$I = \frac{V}{R}$$

If R is the unknown quantity, the formula must again be rearranged.

$$I \times R = E \quad \text{Divide both sides by } I.$$

$$\frac{I \times R}{I} = \frac{E}{I}$$

$$R = \frac{E}{I}$$

These are the three forms of Ohm's Law. Each of the quantities can be solved for using one of the forms.

EXAMPLE 1:   A PV panel supplies 24 volts to a resistance of 8 ohms. What current will flow?

We are given: $E = 24$ volts and $R = 8$ ohms. We are solving for I. The formula used to solve for I is:

$$I = \frac{E}{R}$$

$$I = \frac{24 \text{ volts}}{8 \text{ ohms}}$$

Estimating we get $\dfrac{20}{10} = 2$. Solving we get

$$I = \frac{24 \text{ volts}}{8 \text{ ohms}} = 3 \text{ amperes}$$

Often more than one resistance is wired into a circuit. When using Ohm's Law, only one value for each of the terms is used. Let us find the one value to use for resistance when there are multiple resistances. Resistances can be wired into a circuit in two distinct ways. The first is end to end. There is only one way for the current to flow, and that is through each of the resistances. This is known as a *series circuit*. The second way is end and end. Here there are multiple ways for current to flow. This is known as a *parallel circuit*. We need to find the value to represent resistances wired in each of these ways. That resistance is called the equivalent resistance (R_T).

For resistances wired in series, the value (R_T) to put into Ohm's Law is determined by

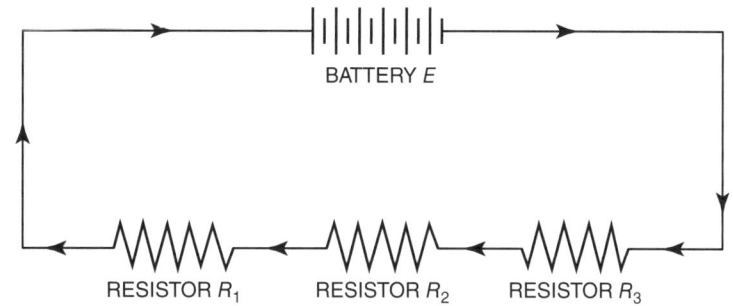

$$R_T = R_1 + R_2 + R_3 + \ldots$$

For resistances wired in parallel, the value (R_T) to put into Ohm's Law is determined by

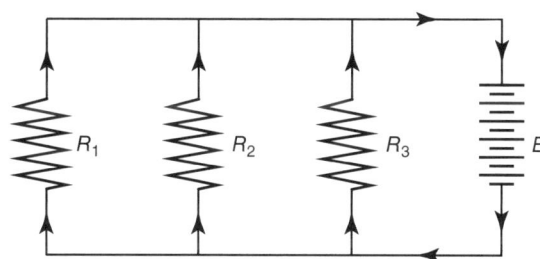

$$R_T = \frac{1}{\dfrac{1}{R_1} + \dfrac{1}{R_2} + \dfrac{1}{R_3} \ldots}$$

It is important to realize that $\frac{1}{\frac{1}{R_1} + \frac{1}{R_2} + \frac{1}{R_3} + \ldots}$ is very different from $R_1 + R_2 + R_3 + \ldots$. One way to remember this is to realize that the more resistances are added in parallel, the smaller R_T becomes, while the more resistances are added in series, the larger R_T becomes.

EXAMPLE 2: Three resistors are wired in parallel. They have resistances of 2 ohms, 3 ohms, and 12 ohms. What is the equivalent resistance for these resistors?

$$R_1 = 2 \text{ ohms} \qquad R_2 = 3 \text{ ohms} \qquad R_3 = 12 \text{ ohms} \qquad R_T = ?$$

$$R_T = \cfrac{1}{\cfrac{1}{R_1} + \cfrac{1}{R_2} + \cfrac{1}{R_3}}$$

$$R_T = \cfrac{1}{\cfrac{1}{2} + \cfrac{1}{3} + \cfrac{1}{12}}$$

The LCD (lowest common denominator) must be found before adding the fractions. Then the fraction must be simplified.

$$R_T = \cfrac{1}{\cfrac{6 + 4 + 1}{12}} = \cfrac{1}{\cfrac{11}{12}} = \frac{12}{11} \text{ ohms} = 1\frac{1}{11} \text{ ohms}$$

Estimating here is difficult because of the complex fractions. There is a rule of thumb when working with resistors in series or parallel. The rule is: For resistors in series, the equivalent resistance is larger than the largest resistor in the series. For resistors in parallel, the equivalent resistance is smaller than the smallest resistor. The answer above is smaller than the smallest resistor's 2 ohms.

Remember:

- Read the problem and make a chart or list of the given quantities and what is asked for.
- Determine the formula that will give the answer asked for.
- Substitute numbers in for the formula letters.
- Be sure the units are correct.

Practical Problems:

- Apply the principles of electrical relationships to the problems in this unit. Round answers to two significant numbers to the right of the decimal point where necessary.

Complete this chart.

	E in volts	I in amps	R in ohms
1	?	3	124
2	240	?	48
3	120	8	?
4	77.7	?	18.5
5	?	11.3	27.6
6	362.71	8.3	?
7	?	0.054	92.93
8	10.504	73.66	?
9	0.192	?	33.72

10. A PV panel is generating 48 volts and produces 8 amps of current. What is the resistance of the circuit?

    _____

11. A low-head hydro generator is putting out 230 volts. The resistance of the circuit is 25 ohms. What current is flowing through the circuit?

    _____

12. A PV panel puts out 6.59 amps of current when connected to a 4.25-ohm-resistance circuit. What is the voltage of the panel?

    _____

13. A heat pump is measured to carry 12 amps of current when running. It is powered from a 230-volt circuit. What is the resistance of the heat pump when it is running?

    _____

14. A wind turbine generates 48 volts of electricity. The resistance of the circuit is 3.25 ohms. What current flows through the circuit?

    _____

15. The wire from a low-head hydro generator to the distribution panel has a resistance of 0.07 ohms. The current flowing through the wire is 11.5 amperes. What is the voltage loss from the generator to the distribution panel?

    _____

16. A pump is used to circulate the water in a solar hot water heater. The pump motor has a coil resistance of 15 ohms and draws 8.33 amperes. What is the voltage rating of the motor?

    _____

17. A wind turbine generates 480 volts and produces 31.6 amperes of current. What is the resistance of the circuit?

_____

18. A farm has 2 wind turbine generators on it. They are wired in parallel with an effective resistance of 1.5 ohms. A current of 17 amps flows through the circuit. What is the voltage for the circuit?

_____

19. When wired in parallel, PV panels produce 8.04 amps of current. If the voltage measures 30.5 volts, what is the resistance in the system?

_____

20. Resistance in cables is measured in ohms per foot. #4/0 cable is rated at 0.000049 ohms per foot. What is the total resistance for 3 lengths of 125 feet of #4/0 cable wired in series?

_____

21. Find the equivalent resistance for a 12-ohm resistor and a 15-ohm resistor wired in series.

_____

22. Determine the equivalent resistance for a 12-ohm resistor and a 15-ohm resistor wired in parallel.

_____

23. A circuit consists of 4 resistors of 0.55 ohms, 14.8 ohms, 0.72 ohms, and 9.5 ohms in series. A current of 8.2 amperes flows through the circuit. What voltage supplies the force for this current?

_____

24. Resistances of 15 ohms, 2 ohms, and 10 ohms are wired in parallel. The voltage of the power supply is 230 volts. What is the current flowing through the main part of the circuit?

_____

25. Three resistances, 0.75 ohms, 3.5 ohms, and 0.6 ohms, are in series. The generator voltage is 30.7 volts. What current flows in the circuit? The two small resistances are from the cables leading to and from the generator. What is the voltage loss in each of the cables?

_____

_____

_____

UNIT 33

Electrical Relationships II

Basic Principles of Electrical Relationships

- Use the electrical formulas found in Section IV of the Appendix.

In the last unit, Ohm's Law and equivalent resistances were studied. There are other electrical formulas that might be encountered or needed in the energy field. One deals with electrical power. Power is the rate at which energy is converted—in our case—to heat or light or motion. The amount of power utilized in an electrical device is given by

$$P = I \times E$$

P is the *power* given in units of watts. I is the current and E is the voltage, which were introduced in the last unit. In the last unit we saw that volts were equal to amperes $\times$ ohms. Here we see that watts are equal to amperes $\times$ volts.

EXAMPLE 1: A PV panel produces 6.5 amperes of current with a voltage of 26.4 volts. What is the power output of the PV panel?

I = 6.5 amperes E = 26.4 volts P = ?

$P = I \times E$

= 6.5 amperes $\times$ 26.4 volts

Estimating gives 7 amps $\times$ 30 volts = 210 watts

Solving = 171.6 watts

Just as the Ohm's Law formula could be rearranged to solve for any of the quantities, the power formula can also be rearranged. This leads to:

$$I = \frac{P}{E}$$

or

$$E = \frac{P}{I}$$

We can go a step further using Ohm's Law and substitute for E or I. Substituting for E leads to:

$$P = I \times E$$

$$= I \times (I \times R)$$

$$= I^2 \times R$$

When current is sent through a wire, there is a power loss in the wire given by this formula. This power loss is energy wasted and not being used efficiently. The power loss can be reduced by reducing the resistance. This is accomplished by increasing the size of the wire, which is one reason a larger wire is used when sending power over a long distance.

Substituting for I leads to:

$$P = I \times E$$

$$= \frac{E}{R} \times E$$

$$= \frac{E^2}{R}$$

In the last unit, resistors in series and in parallel were studied. For the series or parallel circuit, we can also look at the voltage or the current. Voltages can be wired in a series or parallel arrangement. Current can flow in a series or parallel arrangement. Some care needs to be taken when making these kinds of circuits.

Voltages wired in parallel should always be the same voltage.

$$E_{Parallel} = E_1 = E_2 = \ldots$$

If not, they will not work properly. Voltages wired in series will add their voltages.

$$E_{Series} = E_1 + E_2 + \ldots$$

When setting up a PV array that has groups of panels wired in parallel, care must be taken to have the same number of panels in each group to get the same voltage from each group. In this case, the panels in each group are wired in series and their voltages add. The groups are wired in parallel and should all have the same voltage.

EXAMPLE 2: Technician Parker has a four-cell flashlight that uses four D-cell batteries. Each cell is rated at 1.5 volts. What voltage should the flashlight bulb be rated for to work properly?

The batteries are put in so that their voltages add—they are put in series. So the voltage produced is E = 1.5 volts + 1.5 volts + 1.5 volts + 1.5 volts = 6 volts. The bulb should be a 6-volt bulb.

When components are wired in series in a circuit, the current flowing through one component is the same as the current flowing through each of the other components. There is only one path in a series circuit, and all the current must flow through each component.

$$I_{Series} = I_1 = I_2 = \ldots$$

When wired in parallel, there are multiple paths for current flow. Individual currents flow through each path. But these paths come together at some point in the circuit. All of the current flows through this common path. Here, the current is the sum of the currents through each of the individual paths.

$$I_{Common} = I_1 + I_2 + \ldots$$

Remember:

- Read the problem and make a chart or list of the given quantities and what is asked for.
- Determine the formula that will give the answer asked for.
- Substitute numbers in for the formula letters.
- Be sure the units are correct.

Practical Problems:

- Apply the principles of electrical relationships to the problems in this unit. Round off answers to two significant numbers to the right of the decimal point if needed.

Fill in the asked-for quantities in the following chart.

	P in watts	I in amps	E in volts	R in ohms
1	1,056	16	?	–
2	2,400	?	25	–
3	?	14	120	–
4	391.53	?	9.3	–
5	?	0.036	8.44	–
6	370.097	63.7	?	–
7	?	6	–	15.7
8	462.978	5.1	–	?
9	158,723.712	–	97.2	?
10	?	72.8	–	26.4
11	?	–	59.4	0.18
12	0.35	–	105	?
13	4.5	62.4	–	?
14	?	–	4.2	0.14

15. A low-head hydro generating unit is producing 4,900 watts of power. (This is 4.9 kilowatts.) The generator's output is 21.3 amperes of current. What is the voltage being created by the generator?   _____

16. A wind turbine generator is producing 950 watts of power at a voltage of 230 volts. What is the current flow in this system?   _____

17. A length of wire is carrying a current of 14.5 amps. This length of wire has a resistance of 0.12 ohms. What is the power loss due to this section of wire?   _____

18. A PV panel is producing an output of 6.2 amperes at a voltage of 52.4 volts. What is the power output of this panel?   _____

19. A low-head hydro generator is producing 120 volts and is supplying a load of 49 ohms resistance.

    A. Determine the power output of the generator.   _____

    B. Determine the current flowing in the circuit.   _____

20. A wind turbine is delivering 8.6 amperes of current at 28 volts.

    A. What is the power output of the turbine generator?   _____

    B. What is the resistance in the circuit?   _____

21. A low-head hydro generator produces 2,500 watts of power at 230 volts. Determine the current flow from the generator.   _____

22. A PV panel produces 240 watts of power delivering 8.3 amperes of current. What is the voltage of the PV panel?   _____

23. A $\frac{1}{2}$-horsepower pump is used to circulate heat exchange coolant for a geothermal system. It operates on 120 volts. What current does the motor draw? Hint: Convert the horsepower to watts.   _____

24. Four PV panels are wired in series as shown. What is the voltage at the output?   _____

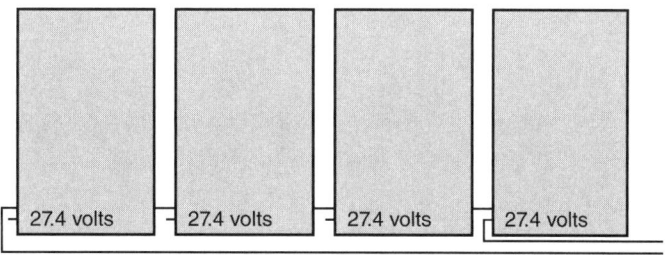

    27.4 volts   27.4 volts   27.4 volts   27.4 volts

25. Three wind turbines are in parallel supplying a distribution box as shown. What current would be leaving the distribution box?   _____

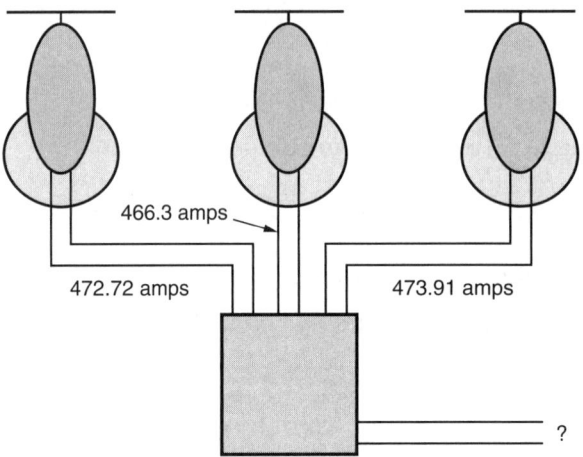

466.3 amps

472.72 amps   473.91 amps

?

UNIT 34

Heat Transport

Basic Principles of Heat Transport

• Study and apply the formulas for heat transport in this unit to the problems that follow.

There are many relationships that can be represented by formulas. In previous units formulas were used to find perimeters, areas, volumes, and electrical quantities. Another important concern with renewable energy is the movement of heat.

When a substance changes temperature, it gains or loses heat. The amount of heat gained or lost can be determined by using the formula

$$Q = m \times c \times \Delta T.$$

Q is the heat gained or lost, m is the amount (mass) of the substance that is being heated or cooled, c is a constant quantity depending on the substance called the specific heat. ΔT is the change in temperature of the substance as a result of the heat being added or taken away. The change is always found by subtracting the initial temperature from the final temperature.

The specific heat is a number depending on the substance and the unit system that is being used. Water has a specific heat of 1 calorie per unit gram per degree Celsius or 1 British thermal unit (Btu) per pound per degree Fahrenheit. This gives the unit of heat to be either calories or Btu. If the heat pump system is using liquid other than water, for example, ethylene glycol so that it would not freeze in very cold weather, the specific heat would be something other than 1. The units would be the same, only the number would change.

EXAMPLE 1:   Two pounds of water increase in temperature 15 degrees. How much heat must be added to the water?

$$m = 2 \text{ pounds} \qquad \Delta T = 15 \text{ degrees Fahrenheit}$$

Since the units are pounds and degrees Fahrenheit, $c = 1$ Btu per pound per degree Fahrenheit.

$$Q = m \times c \times \Delta T$$

$$= 2 \text{ pounds} \times 1 \frac{\text{Btu}}{\text{pound} \times \text{degree Fahrenheit}} \times 15 \text{ degrees Fahrenheit}$$

Estimating:

$$Q = 2 \times 1 \times 10 = 20$$

Solving

$$Q = 2 \text{ pounds} \times 1 \frac{\text{Btu}}{\text{pound} \times \text{degree Fahrenheit}} \times 15 \text{ degrees Fahrenheit} = 30 \text{ Btu}$$

Heat was gained by the water. If it were lost, the number might be the same, but a minus sign would be with the number. The sign comes from subtracting the initial temperature from the final temperature. A minus sign will appear when the object is cooling down.

A second formula for heat transport is one that describes how much heat moves through a wall such as the wall of a pipe in a heat exchange bed in a time period. This formula is

$$Q = U \times A \times \Delta T$$

where A is the area of the wall the heat is passing through and U is the heat transfer coefficient telling us how easily heat can pass through the wall. Insulation makes it harder for the heat to pass through the wall, so insulation affects U. U has units of Btu/hour square feet degrees Fahrenheit.

U and c are constant values that are usually given to you or can be found in tables.

EXAMPLE 2:   A horizontal heat exchange bed for a ground source heat system has 480 sq ft of tubing that has a U of 8 Btu/hour sq ft degrees Fahrenheit. If glycol enters the bed with a temperature of 39 degrees Fahrenheit and leaves with a temperature of 51 degrees Fahrenheit, how much heat is being transferred to the system each hour?

$$U = 8 \text{ Btu/hour sq ft degrees Fahrenheit} \qquad A = 480 \text{ sq ft}$$

$$\Delta T = (51 - 39) = 12 \text{ degrees Fahrenheit}$$

$$Q = U \times A \times \Delta T$$

Estimating:

$$Q = 10 \times 500 \times 10 = 50,000$$

Solving:

$$Q = 8 \text{ Btu/hour sq ft degrees Fahrenheit} \times 480 \text{ sq ft} \times 12 \text{ degrees Fahrenheit}$$

$$= 46,080 \text{ Btu/hr}$$

An important point to remember is that the heat lost by one fluid must equal the heat gained by the other fluid in a heat exchanger. So the Q_{gained} by liquid 1 = Q_{lost} by liquid 2.

Practical Problems:

• Apply the principles of heat transport to the problems in this unit.

1. A horizontal heat exchange bed is gaining heat for a heat pump in the winter. The circulating water is flowing at a rate of 9 gallons per minute (72 pounds per minute) and enters the heat exchange bed at a temperature of 47°F. The water leaves the bed with a temperature of 52°F. How much heat is gained by the water in 1 minute? _____

2. A vertical well has a water flow of 50 pounds per minute. Water enters the well at a temperature of 82°F and leaves the well at a temperature of 63°F. How much heat would be given to the well in 1 minute? _____

3. A solar collector has water flowing through it. On a sunny day, the collector has water entering the collector at 79°F and leaving the collector at 107°F. The water is removing 1,568 Btu of heat each minute. What is the flow of the water through the collector? _____

4. A larger collector would be able to put more Btu into the system. It would also have a larger flow. With the same ΔT as the previous problem, what flow would be needed to remove 2,352 Btu of heat?   _____

5. The collector in problem 3 is operating on a partly sunny day and only collects 1,176 Btu of heat each minute. If the flow remained the same as in problem 3, what was the temperature change of the water?   _____

6. A 50% mixture of ethylene glycol and water can be used in solar collectors and will not freeze until the temperature drops to about −40°F. The specific heat of a 50% mixture of ethylene glycol and water is 0.81 Btu per pound per degree Fahrenheit. A 56 pound per minute flow of this mixture through a solar collector is removing 1,568 Btu of heat each minute. What size temperature change will this system have?   _____

7. Solar insolation is the amount of heat contained in the sunlight as it strikes a surface. During one month at Mytown, the solar insolation is 80.76 Btu per minute per square foot. How much heat is collected each minute by the solar collector shown?   _____

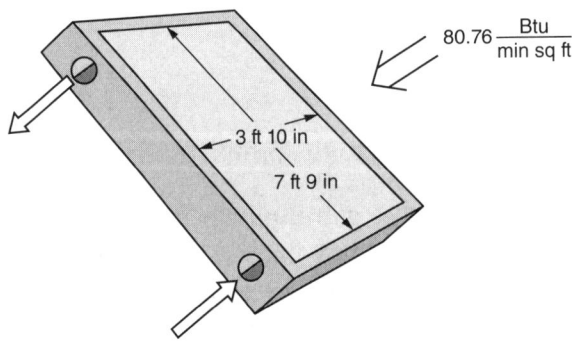

8. Because of cloudiness, the sun in Mytown shines the equivalent of 4.4 hours one day. What is the total heat collected by the solar collector in problem 7 for that day?   _____

9. With the sun shining in Mytown as described in problem 7, the solar collector has 77 pounds of water flowing through the collector each minute. What is the temperature change of the water in the collector?   _____

10. The solar collector on a house in Mytown is fixed, so it does not change angles. The sun does change angles, so it does not strike the collector at the best angle all of the time. In the spring, the collector only collects 92% of the normal amount of solar insolation. With the same flow through the collector described in problem 9, what is the temperature change for the coolant flowing through the collector? _____

11. A hot water system has a collector that is 3 feet by 5 feet in size and is placed at a site that receives 74.3 Btu per minute per square foot. But the system has losses and is only 94% efficient. What is the useful heat available to be used each minute? _____

12. A geothermal well system is shown. How much heat is rejected to the well, if the fluid used is water? _____

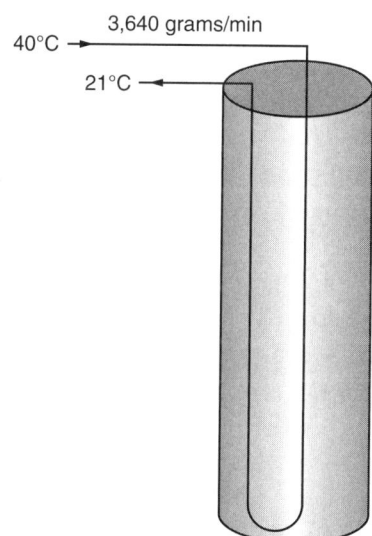

3,640 grams/min

40°C

21°C

13. Technician Carson was called to a hot water system that was not performing as it should. Carson found that instead of moving 65,000 calories of heat, it is only moving 45,500 calories. The collector is only collecting what percent of the available heat? (Carson found that the owner had painted the collector plate white, not black.) _____

14. The owner of the white solar collector did not want to change the color of the collector, but wanted to have the water leaving the collector to have a 19°C temperature increase. What water flow would be needed for this temperature increase?   _____

15. A solar hot water system has a collector providing 55 pounds of water per minute with a 26°F increase in temperature. How much heat is being moved each minute?   _____

16. The solar hot water system in problem 15 is used to heat an 80-gallon (640 pound) water tank. How much will the temperature be increased each minute the solar collector is producing at the rate given in problem 15?   _____

17. What would be the temperature increase in a 100-gallon (800 pound) tank using the collection system from problem 15?   _____

18. A solar collector is mounted on a roof that points to the southwest rather than due south. This results in a reduction in solar insolation to 92% of the maximum amount of 76.5 Btu per minute per square foot. What is the amount of heat collected each minute in a 4 foot by 6 foot collector?   _____

19. A home geothermal heating/cooling system has a 30% ethylene glycol mixture in water, producing a specific heat of 0.89 Btu per pound per degree Fahrenheit. The warm water enters a heat exchange well at a temperature of 92°F and leaves the well at a temperature of 70°F. With a flow rate of 75 pounds of coolant per minute, how much heat is rejected to the well?   _____

20. In the winter the home geothermal heating/cooling system in problem 19 is used to heat the home. Now the coolant is entering the well at a temperature of 52°F and leaves the well at a temperature of 67°F. How much heat is extracted from the well?   _____

21. Technician Makena installed a solar hot water system in her house. Her 80-gallon storage tank is insulated so that heat loss from the tank is reduced. The surface area of the tank is 25.7 square feet. The insulation gives the tank a U value of 7 Btu/hour square feet degrees

Fahrenheit. With water temperature at 110°F and the temperature of the room at 71°F, how much heat is lost to the room each hour?

22. Technician Makena decides to add another layer of insulation to the storage tank. Covering the tank with a layer of R-19 insulation reduces the U value to 5 Btu/hour square feet degrees Fahrenheit. How much less heat is lost from the tank each minute?

23. A horizontal heat exchange bed is composed of 1,250 feet of 1.5-inch-diameter high density polyethylene (HDPE) pipe. It has a surface area of 490.9 square feet and a heat transfer coefficient of 0.057 Btu per minute per square foot per °F. If the ΔT across the tube is 23°F, how much heat is transferred each minute?

24. Copper tubing can be used in the heat exchange bed instead of HDPE tubing. For a thick wall tube, U increases from 2.7 Btu per hour per square foot per °F to 19 per hour per square foot per °F. If the HDPE tubing had to be 1,100 feet long, how long would the same diameter copper tubing need to be to transfer the same amount of heat?

25. A layer of dirt built up on the inside of the tubing of a horizontal heat exchange bed for a geothermal system. This results in the heat transfer coefficient decreasing from 0.0587 Btu per minute per square foot per degree Fahrenheit to 0.0513 Btu per minute per square foot per degree Fahrenheit. If the tubing surface area is 294.5 square feet and the temperature difference is 15°F, how much less heat is transferred each minute?

UNIT 35

Energy

Basic Principles of Energy

- Study the formulas related to energy given and described below.

Energy is the ability to do work. When work is done, energy changes form. There are many different forms of energy—heat energy, light energy, potential energy, kinetic energy, and chemical energy to name a few. One of the more convenient forms of energy is electrical energy. The wind turbines and hydro turbines generate electrical energy. They do this by converting a different form of energy to the electrical energy. The wind turbines convert energy contained in the wind to electrical energy. The hydro turbines convert water energy to electrical energy. PV cells convert light energy to electrical energy. Batteries convert chemical energy to electrical energy. Electrical energy is found by multiplying electrical power times time.

Utilizing renewable energy sources, electrical energy is produced. Heat energy can also be produced using solar collectors. In all of these cases, energy changes form. There is a rule of nature that energy is not created or destroyed, it only changes form. The amount of work that can be done depends upon the amount of energy that is available. This is one reason it is important to determine the energy available. There are a number of formulas that allow us to find the amount of energy available.

Potential energy is energy available due to the position of a substance. Examples of this include a coiled spring, a lifted weight, or water level built up behind a dam. When things are lifted, the potential energy can be determined by:

$$PE = W \times h = m \times g \times h$$

where W is the weight of the material and h is the height that it is raised. Water held in a dam has potential energy, and the h there is the height that the water will fall. The m is for the mass of the material (this is related to but not equal to the weight) and g is a constant value called the acceleration due to gravity. From the formulas, we can see that the weight of an object and its mass are related using the acceleration due to gravity. The reason to go through all of this is that when the metric system is used, everything is talked about in terms of mass and when the English system is used weight is what is used and discussed.

EXAMPLE 1: How much potential energy does a pound of water have that is raised 3 feet?

$$W = 1 \text{ pound} \qquad h = 3 \text{ feet}$$

$$PE = W \times h$$

$$= 1 \text{ pound} \times 3 \text{ feet}$$

$$= 3 \text{ foot-pounds}$$

Kinetic energy is energy as a result of material moving. This could be water flowing or wind blowing. The kinetic energy can be found by:

$$KE = \frac{1}{2} \times m \times v^2$$

Where m is the mass of the material and v is the speed of the material moving.

EXAMPLE 2: How much energy does 2 kilograms of water have when flowing at 3 meters per second?

$$KE = \frac{1}{2} \times m \times v^2$$

$$= \frac{1}{2} \times 2 \text{ kilograms} \times \left(3 \, \frac{\text{meters}}{\text{second}} \right)^2$$

$$= 1 \text{ kilogram} \times 9 \, \frac{\text{meter squared}}{\text{second squared}}$$

$$= 9 \, \frac{\text{kilogram meter squared}}{\text{second squared}} = 9 \text{ joules}$$

The units for energy are foot-pounds in the English system and joules in the metric system. The joule is the same thing as (another name for) a kilogram meter squared per second squared. A kilogram is equal to 1,000 grams. Another unit for energy in the English system is the British thermal unit. Electrical energy is watt-hours or kilowatt-hours. (The units are getting complicated. Care must be taken to keep the units correct.)

Practical Problems:

- Apply the principles of energy to the problems in this unit. Round answers to two significant numbers after the decimal point where necessary.

1. How much energy does 5 pounds of water have as a result of raising it 12 feet?   _____

2. Water is moving at a speed of 4 feet per second. How much kinetic energy is possessed by a mass of water of 3 slugs? (Slugs is the unit of mass in the English unit system.)   _____

3. Air is blowing at 5.36 meters per second. How much kinetic energy would each kilogram of air possess?   _____

4. A low-head hydro generator is producing 3 kilowatts of power. How much electrical energy is created in 24 hours?   _____

5. A hydro dam is located 20 feet above the turbine. A decision is made to raise the dam breast 4 feet. How much more energy will each gallon (8 pounds) of water have with the higher dam?   _____

6. The more energetic water from the higher dam in problem 5 drives a turbine generator. How big an increase in generator electrical output would each gallon of water produce over the old dam height?   _____

7. Solar insolation one day at Mytown is 79.8 Btu per hour per square foot. A solar collector measuring 3.5 feet by 5.5 feet is set to maximize the collection. How much energy is collected in an hour?   _____

8. How much potential energy does 1 pound of water have with the
   system shown?   _____

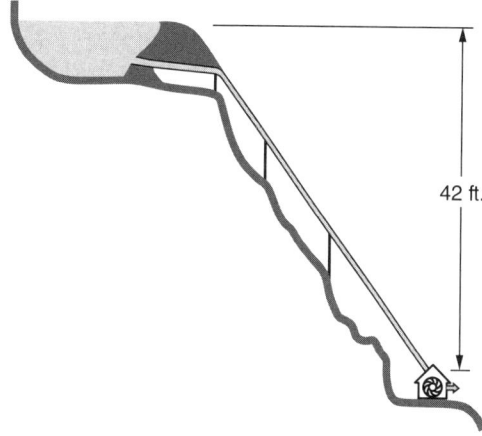

42 ft.

9. A dry period causes the water level in the dam in problem 8 to be
   6 inches below the top of the dam. What is the energy that is available
   for the generator?   _____

10. The hydro generator in problem 9 is 93% efficient in converting to
    electrical energy. How much energy is produced per minute if the
    penstock provides 2,200 gallons per minute?   _____

11. How much electrical energy would be generated by this system in one day? _____

12. A solar collector is shown. How much energy is taken in by the
    collector in one hour?   _____

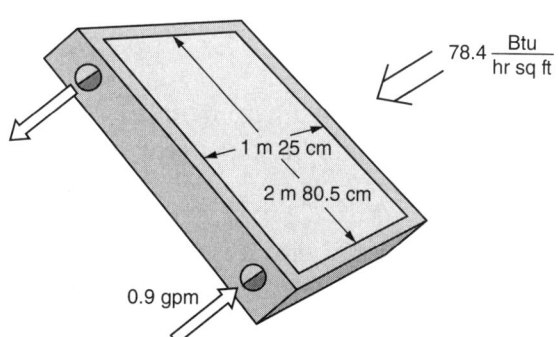

$78.4 \frac{Btu}{hr\ sq\ ft}$

1 m 25 cm

2 m 80.5 cm

0.9 gpm

13. The efficiency of the collector in problem 12 is 85%. How much energy is converted to useful heat? _____

14. Water is shown flowing through the collector in problem 12. What is the increase in temperature of the water as it passes through the collector? _____

15. A wind turbine site has wind blowing at 6.26 meters per second. For conditions at Yourtown, the mass of air passing (interacting with) the rotor each second is 80.9 kilograms. How much kinetic energy does this air contain? _____

16. A reduction in wind speed to 5.81 meters per second will result in a mass of 75.1 kilograms of air interacting with the turbine rotor. What is the loss in kinetic energy as a result of this decrease in wind speed from problem 15? _____

17. The efficiency of the wind turbine in problem 15 is 48%. How much electrical energy is produced? _____

18. The electrical energy produced in problem 17 has units of joules. A joule is the same thing as a watt-second. Since the wind speed has that mass of wind moving each second, the electrical energy is producing that much power each second. So the turbine is producing that number of watts each second. Thus, the power is that number of watts. Using the formula for power, if the turbine is producing this power at 240 volts voltage, what current is created? The power is transmitted through #6 gauge wires that have a total resistance of 0.198 ohms. _____ What is the power lost in transmitting the current through these wires? _____

19. What is the total energy lost just in these wires in one day? Hint: Multiply the power times the time. _____

20. Technician Makena suggested that #4/0 wire be used. It has a total resistance of 0.025 ohms. What would be the energy lost in one day using this wire? _____

21. A PV solar panel is 1,650 cm by 1,000 cm. It is oriented to receive a solar insolation of 79.35 Btu per minute per square foot. How much energy does the panel receive in 1 second? _____

22. The efficiency of the panel in problem 21 is 16.7%. How much electrical energy is produced in that 1 second?

_____

23. The PV panel in problem 21 produces a voltage of 29 volts. What current flows from this panel?

_____

24. A transformer uses the conservation of energy. The amount of electrical energy entering the transformer is the amount of electrical energy leaving the transformer. Transformers change the voltage of the electricity. As a result, current also changes, because the product must remain a constant. The outgoing voltage is 240 volts. What is the outgoing current?

_____

_____

25. By increasing the voltage in the previous problem, less energy is wasted in the wires. How much energy is saved each second in a 2,000-meter length of #4/0 wire which has a resistance of 0.098 ohms?

_____

UNIT 36

Trigonometric Functions

Basic Principles of Trigonometric Functions

- Use the table of values for trigonometric functions found in Section V of the Appendix.

Triangles are three-sided figures. The sides form three angles. The sizes of the three angles always add up to 180°. Triangles are named depending on the size of the angles. If the triangle has one angle that is exactly 90° (a right angle), the triangle is named a right triangle. When two triangles have the same angles, the lengths of the sides will have the same ratios. This property can be used to determine lengths of sides or angles of triangles.

The sides of a right triangle are given specific names. These names are used to help develop a table for ratios of the lengths of sides of a right triangle. These ratios of side lengths are called trigonometric functions.

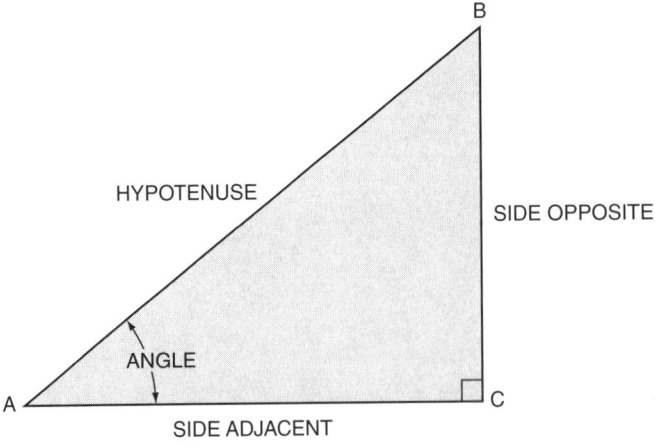

The trigonometric functions and their ratios are:

$$\text{SINE} \quad \text{Sin} = \frac{\text{Side opposite}}{\text{Hypotenuse}} \qquad \text{COSINE} \quad \text{Cos} = \frac{\text{Side adjacent}}{\text{Hypotenuse}}$$

$$\text{TANGENT} \quad \text{Tan} = \frac{\text{Side opposite}}{\text{Side adjacent}}$$

The names of the sides depend upon which angle is being talked about. When a right triangle is drawn, the right angle is always labeled as angle C. The side of the triangle that does not make up the angle is called the hypotenuse. Looking at one of the other angles, it can be seen that the hypotenuse is one of the sides making up that angle. The other side making up the angle is termed the adjacent side. The side that is not making up the angle is termed the opposite side. Choosing the other angle results in the two sides (adjacent and opposite) exchanging their names.

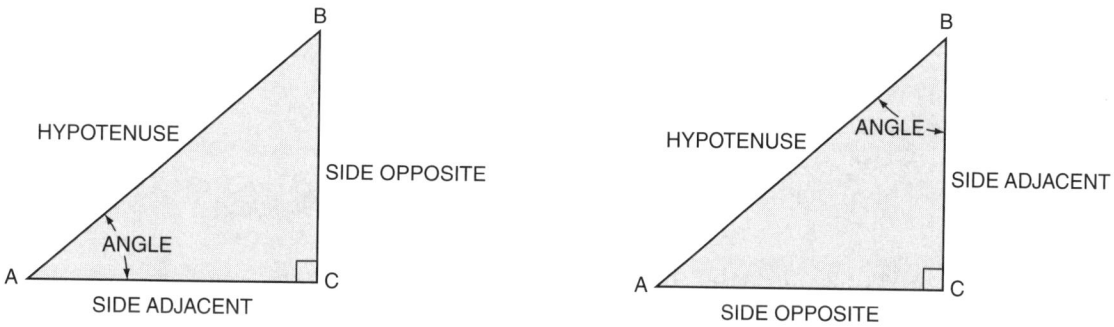

The definitions of the trigonometric functions stay the same. The lengths of the sides will be different for changing angles, so the ratio will change for each angle. The lengths of the sides will change for different size triangles, but the ratio of the lengths of the sides will stay the same. Each trigonometric function is written as a fraction. This means that there are three quantities involved. By knowing any two, the third can be found. As an example, use the sin.

$$\text{Sin} = \frac{\text{Side opposite}}{\text{Hypotenuse}} \qquad \text{Hypotenuse} = \frac{\text{Side opposite}}{\text{sin}} \qquad \text{Side opposite} = \text{sin} \times \text{Hypotenuse}$$

Similar relations can be written for cos and tan. These relations are used to find the other properties of any triangle. The properties would be the lengths of the sides and the size of the angles. When finding these properties, it is best to draw a right triangle and put in the values that are given. This will help you visualize what needs to be found and which are the terms involved.

EXAMPLE 1:   Determine the size of angle A.

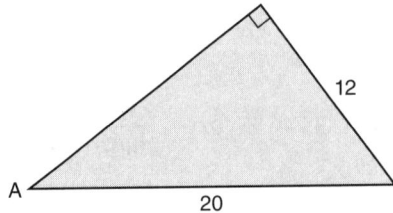

For angle A the two given sides are the hypotenuse (opposite the right angle) and the opposite side.

$$\text{Hypotenuse} = 20 \qquad \text{Side opposite} = 12$$

The sin is the trigonometric function to use to solve this.

$$\text{Sin} = \frac{\text{Side opposite}}{\text{Hypotenuse}}$$

$$= \frac{12}{20}$$

$$= 0.6$$

From the table we see that this is the sin for angle 37°. So angle A = 37°.

EXAMPLE 2:   Find the length of side a.

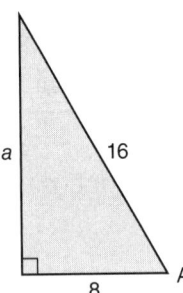

For angle A, we are given the adjacent side. The hypotenuse is also given.

$$\text{Hypotenuse} = 16 \qquad \text{Side adjacent} = 8$$

The cos is the trigonometric function to use.

$$\text{Cos A} = \frac{\text{Side adjacent}}{\text{Hypotenuse}} = \frac{8}{16} = 0.5$$

From our trigonometric tables, the angle whose cos is 0.5 is 60°. We know the angle, but we are after the length of side *a*, which is the opposite side to angle A.

$$\text{Side opposite} = \text{Sin} \times \text{Hypotenuse}$$

$$\text{Side opposite} = \text{Sin } 60° \times 16$$

$$= 0.866 \times 16$$

$$= 13.856$$

Practical Problems:

- Apply the principles of trigonometric functions to the problems in this unit.
- Use the table of trigonometric functions found in the Appendix.

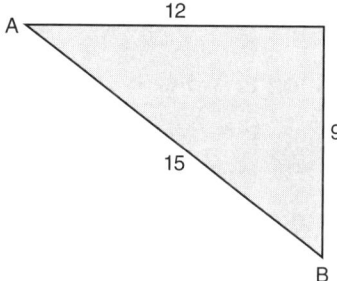

Use the above figure for problems 1−3.

1. Determine the sin of angle A (round to two decimal places). _____

2. Determine the cos of angle A (round to two decimal places). _____

3. Determine the tan of angle B (round to two decimal places). _____

4. Give the value for sin 45° (round to two decimal places). _____

5. Find the value for cos 63° (round to two decimal places).   _____

6. What is the tan 20°? (Round to two decimal places.)   _____

For problems 7 through 14, determine the requested values in the table below:

	Angle	Opposite	Adjacent	Hypotenuse
7.	37°	7	–	?
8.	60°	–	5	?
9.	50°	10	?	–
10.	?	5	–	8
11.	?	4	10	–
12.	?	–	6	16
13.	45°	–	?	20
14.	20°	?	–	15

15. A penstock makes the bend shown. What is the angle made at the bend?   _____

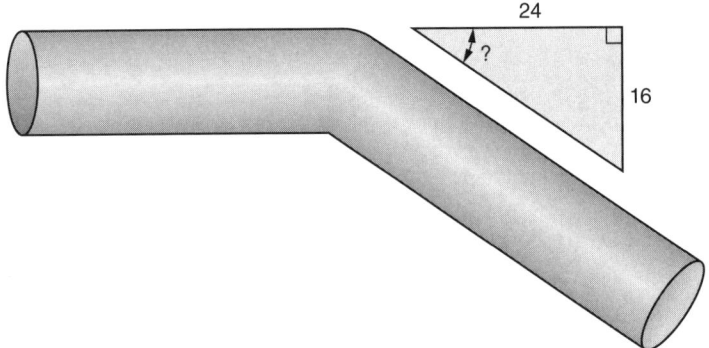

16. Determine the angle of the roof. _____

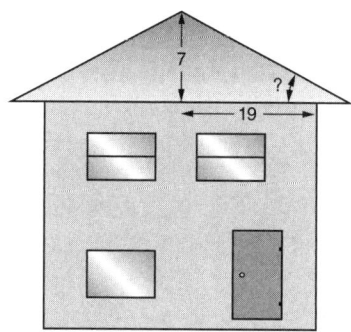

17. Dirt is used to support the breast of a low-head hydro dam. What is the angle of the face of the dirt support? _____

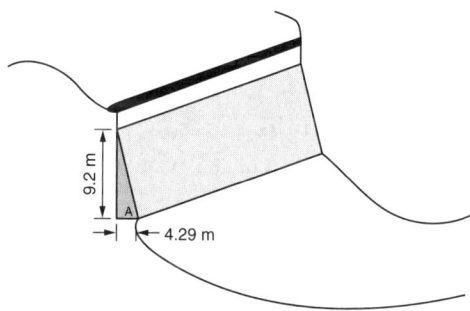

18. How long must the pipe be to reach the geothermal heat exchange horizontal bed? _____

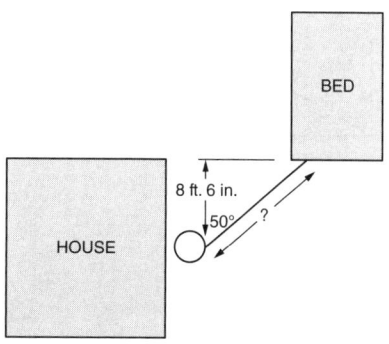

19. How long must the pipe be coming off the roof for the solar hot water
     system shown?   _____

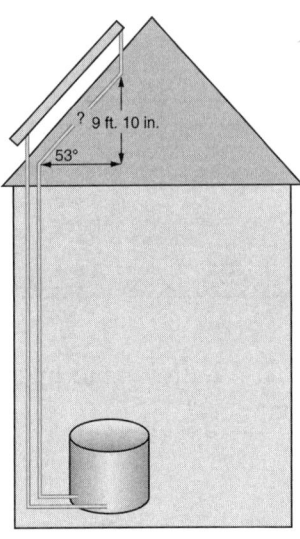

20. A hillside makes an angle of 15° with the horizon. How far down the
     hill from a starting point would the vertical drop be 25 meters?   _____

21. How long is the vertical support for the solar hot water panel shown?   _____

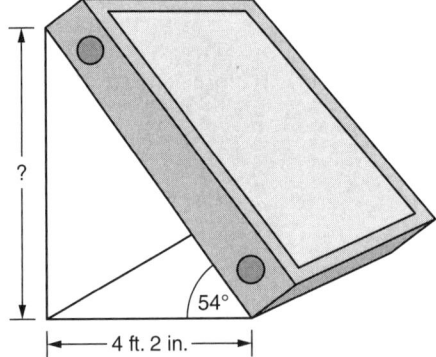

22. A PV panel has an angled support to properly orient it. How far up the panel is the base attached?

_____

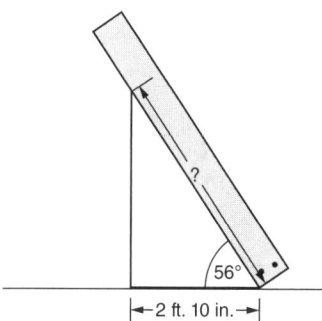

56°

|←2 ft. 10 in.→|

23. Determine the vertical drop of the penstock shown.

_____

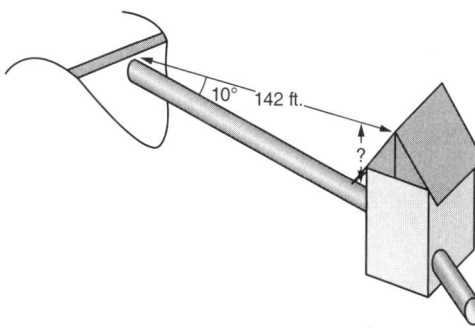

10°   142 ft.

?

24. A wind tower has a support in the shape of a right triangle. How high up the tower does the top of the support reach?

_____

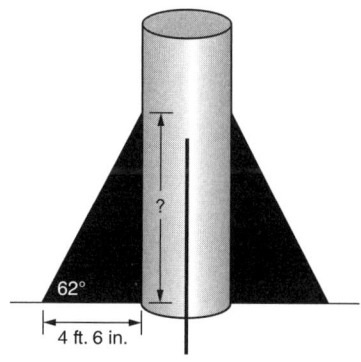

?

62°

|←4 ft. 6 in.→|

25. A support holds a PV panel on a roof. What is the separation of the top
     of the panel from the roof?   _____

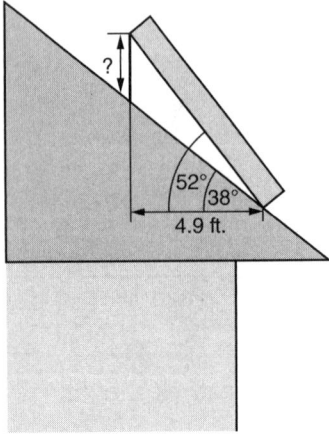

SECTION

Graphs

UNIT 37

Graphs and Graphing

Basic Principles of Graphs and Graphing

Graphs are ways of displaying quantities of information. A graph can make the relationship between two quantities visual. These two quantities are termed variables. The variables are displayed on the two axes of the graph. Each axis has a uniform scale and units displayed. It is important to remember that the two scales do not have to be the same scale, but each scale has to be uniform.

Additional information can be displayed on a single graph, but that would require additional axes or additional lines (curves) to be added to the graph. If that is done, care must be taken to use the correct axis when obtaining information from the graph. Obtaining the information from the graph is called reading the graph.

There are times when information displayed in a table would be better displayed on a graph. So a graph needs to be created. When creating a graph, decide what is to be displayed. Next, determine the extent of the values. The whole graph should be filled with the values that are being displayed, not just a small portion of the graph.

NOTE: A graph may have many data points plotted. Some of them may be plotted a distance from the scale that would give its value. When reading such a graph, make sure you are following the proper line to the correct scale to determine its value. A ruler or a straightedge can help when reading the graph.

The scale should include the units for that axis. When creating a graph, that information should be included. The scale should also be chosen so that the graph fills the entire page, not just one small portion of it. That would make reading the graph easier.

There are different types of graphs. They display the data in different ways. One type is a bar graph. A bar graph displays the data as a solid bar. The length of the bar is related to the size of the quantity being displayed by the graph. Each bar represents a different data point. A different type of graph is the line graph. The line connects the different data points.

EXAMPLE 1:   A bar graph shows the overtime hours for the Good for the Planet Energy Company for last year.

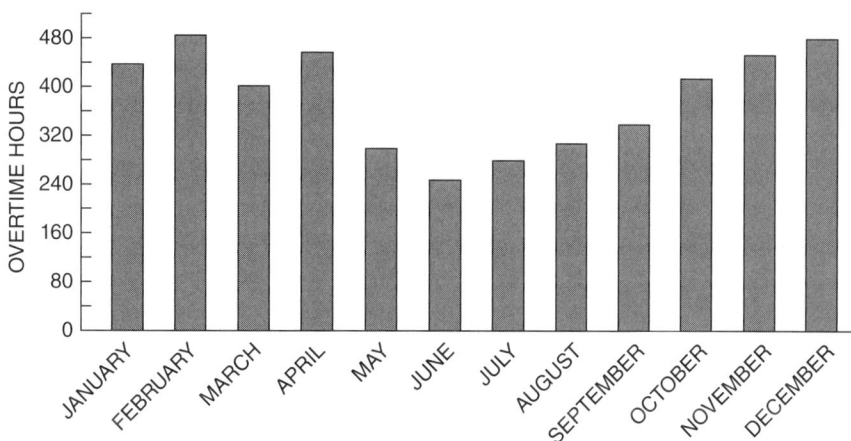

If the hours are spread evenly over the month, can an extra person be hired full time by the company and be kept busy? Can two be hired full time?

A 40-hour week would equate to a 160-hour month. The least overtime month (June) had about 240 hours, so a person could easily be hired full time. For two people to be hired full time, the least amount of overtime would require at least 320 hours. The least amount of overtime is about 240 hours, so a second person would not be kept busy all the time. The second full-time person should not be hired.

A line graph is a figure that has a line connecting data points. Any point on the line shows the relation between the two variables. You get the value for each point by moving vertically up or down from the horizontal axis to the line or horizontally left or right from the vertical axis to the line. At the line, one moves at right angles to the direction you moved to get to the line to read the other variable off of the scale on the axis. When using line graphs, be sure to move only absolutely vertical or absolutely horizontal to be accurate getting data from the graph or putting data on the graph.

Practical Problems:

- Apply the principles of graphing to the problems in this unit.

1. The following service calls were recorded by the We're Green Energy Company for the last year: January 132, February 245, March 142, April 87, May 125, June 254, July 148, August 229, September 237, October 124, November 140, and December 238. Make a bar graph of this data. _____

2. If a technician can, on the average, make 2 service calls per day and the month has an average of 20 work days, make another graph of the number of men needed each month for the data in problem 1. _____

3. Running a "Get ready for the summer" special will increase the number of calls by 25% during the month the special is run. If run in April for the We're Green Company, how many more technicians would be needed that month? _____

4. Measurements were made on the amount of solar collector output for various orientations. For a particular location, a collector is oriented at a 35° angle with the horizon. The following percentages of the maximum amount of solar energy on the collector are recorded: 90° West – 84%, 75° West – 87%, 60° West – 91%, 45° West – 95%, 30° West – 98%, 15° West – 99%, South – 100%, 15° East – 99%, 30° East – 98%, 45° East – 96%, 60° East – 93%, 75° East – 89%, and 90° East – 85%. Create a line graph from this data. _____

5. How much variation in the orientation of the collector can there be and still capture at least 90% of the incoming solar energy? _____

6. How much variation in the orientation of the collector can there be and still capture at least 95% of the incoming solar energy? _____

The attached graph is used for problems 7–9.

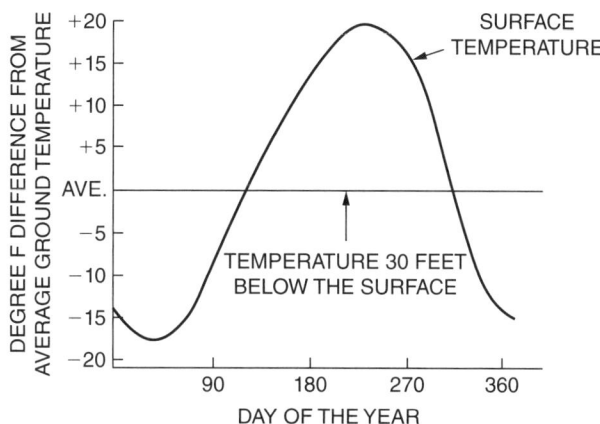

7. The graph shows the ground temperature at the surface and at 30 feet below the surface for a plot of land. If the temperature 30 feet below the surface is a constant 58°F, what are the maximum and minimum temperatures at the surface?

8. If the temperature 15 feet below the surface is half the variation, how many days of the year would the ground temperature be within 5°F of the constant value?

9. If the heat pump used for a system in problem 7 area is air cooled and cannot extract heat from the air if the air temperature drops below 42°F, how many days will the system not work?

The attached graph is used for problems 10–13 to help determine if a wind turbine would be able to create electrical power. _____

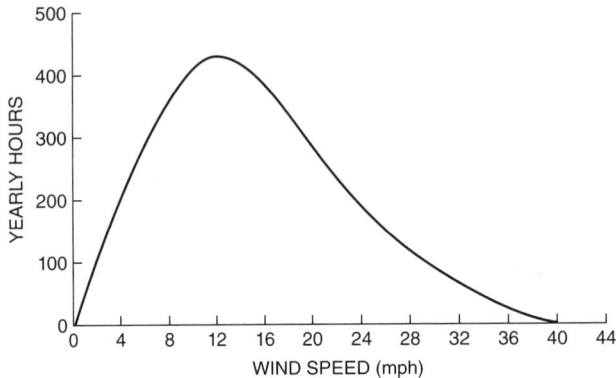

10. A wind turbine will not produce power if the wind has a speed less than 4 mph. How many hours a year will this turbine not be producing power due to 3 mph wind?   _____

11. To prevent the destruction of the turbine, it automatically locks up at wind speeds greater than 34 mph. How many hours a year will the wind be blowing 35 and 36 mph?   _____

12. What range of wind speeds occur for greater than 300 hours a year?   _____

13. What wind speed is experienced most often, and how often does that speed occur?   _____

Problems 14–16 deal with the attached graph. The following graph relates power output to wind speed for a certain turbine generator.

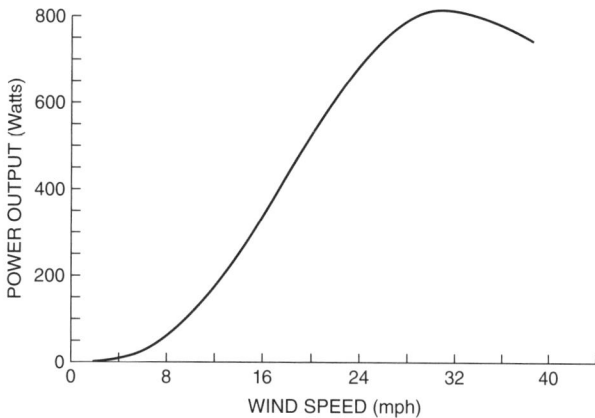

14. What wind speed produces the most power? _____

15. If the wind turbine has a cutoff switch that will not allow the turbine blades to turn if the wind speed is less than 8 mph, what is the lowest power output from the generator? _____

16. Any wind speed greater than 20 mph will produce at least how much power output? _____

Combine the information from the two previous graphs to answer questions 17–20.

17. How many hours a year was the maximum power output experienced? _____

18. What power output was produced from the most often experienced wind speed? _____

19. At what power output will the turbine lock out due to too much wind speed? _____

20. The power output of 800 watts can be produced by two different wind speeds. What is the total yearly number of hours with an output of 800 watts? _____

21. The efficiency of a solar collector depends upon the flow rate of the coolant through the collector. A typical collector gives the following data: 0.01 gpm/sq ft – 46%, 0.03 gpm/sq ft – 49%, 0.05 gpm/sq ft – 50%, 0.07 gpm/sq ft – 50.5%. Make a line graph of the data.

_____

22. What would the efficiency of the collector in problem 21 be if the coolant flow was 0.02 gpm/sq ft?

_____

23. What would the total flow through a 4 ft × 6 ft solar collector be to be at least 49.5% efficient?

_____

The following graph relates the size of the flow of water and the height that the water needs to fall in order to generate 500 watts by a low-head hydro system. Use this graph for problems 24 and 25.

_____

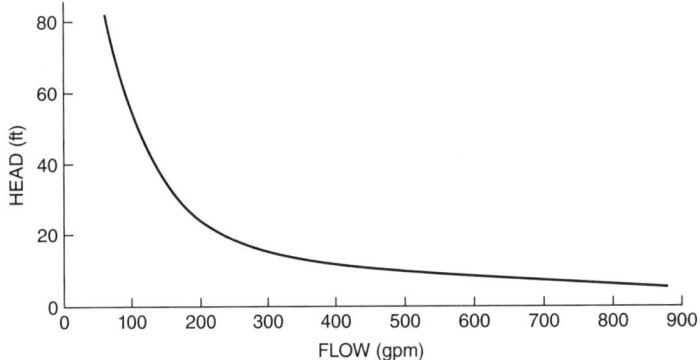

24. Determine the minimum flow to produce 500 watts of power with a water head of 40 ft.

_____

25. The stream that is being used to supply water for this hydro generator has a steady flow of 100 gpm. What is the minimum height above the generator to have the supply generator built?

_____

SECTION

Bills

UNIT 38

Estimates and Billing

Basic Principles of Estimates and Billing

A bill can be filled out for a number of reasons. One is to give a record of what inventory was used. Another is to give a record of how much time individual workers spent on various jobs. A third reason is because a customer has asked for it. The customer then has a record of what equipment was replaced or repaired. There are times when potential customers would ask for a written estimate for the cost of a job.

It is important to be able to correctly write a bill to be able to give it to the customer. To make it understandable to everyone using the bill, the bill needs to be filled out clearly, completely, and correctly. Write a clear description of the pieces used, the labor involved, or the basic charge made. Put down the price per piece when multiple items are used. This should go in the price per column. The price per item times the number of items is the amount charged and should go in the price column.

If the bill is giving an estimated cost, clearly mark **estimate** on the bill. It is also a good practice to write how long the customer has to decide if he or she wants the job done for that price.

When filling out bills and estimates, it is important to fill out as much as possible and to be as clear as possible. The aim is to have someone else read the sheet and understand it completely. Take time to be sure that nothing is uncertain or unclear.

EXAMPLE 1:   Technician Brett makes an emergency call to a home that has a leak in the solar hot water system.

The charges are:

Emergency call $55.00

2 ft piece of 1/2" copper piping @ $2.79/ft $5.58

Labor   2 hours @ $35.00/hr ……………...……..$70.00

Create the bill for this call.

<table>
<tr><td colspan="5" align="center">**WE HAVE IT SUPPLY COMPANY**</td></tr>
<tr><td>**Qty**</td><td align="center">**ITEM**</td><td></td><td>**Price Per**</td><td>**PRICE**</td></tr>
<tr><td>1</td><td colspan="2">Emergency callout</td><td>$55.00</td><td>$55.00</td></tr>
<tr><td>2 ft.</td><td colspan="2">1/2" capper tubing</td><td>$2.79 / ft.</td><td>5.58</td></tr>
<tr><td>2 hr</td><td colspan="2">Labor</td><td>$35.00 / hr</td><td>70.00</td></tr>
<tr><td></td><td></td><td></td><td></td><td></td></tr>
<tr><td></td><td></td><td></td><td></td><td></td></tr>
<tr><td></td><td></td><td></td><td></td><td></td></tr>
<tr><td></td><td></td><td></td><td></td><td></td></tr>
<tr><td></td><td></td><td></td><td></td><td></td></tr>
<tr><td></td><td></td><td></td><td></td><td></td></tr>
<tr><td></td><td></td><td></td><td></td><td></td></tr>
<tr><td></td><td></td><td colspan="2" align="right">**TOTAL**</td><td>$130.58</td></tr>
</table>

Practical Problems:

• Apply the principles of estimates and bills to the problems in this unit.

1. Write an estimate for the installation of a 4,000-watt PV system. The panels cost $23,680.00. Electrical equipment and wiring run $2,270.00. It will take 40 man-hours to install the system at $50.00 per hour.   _____

No.	WE HAVE IT SUPPLY COMPANY Ourtown. U.S.A.	Date:	
Qty	**ITEM**	**Price Per**	**PRICE**
		TOTAL	

2. Create an estimate for annual service for a 2 kW wind turbine. The service call is $40.00. There will be 3 hours of labor at $55.00 per hour. A new gasket and lubrication will cost $62.50.

_____

Qty	No.	**KEEP IT GREEN AND CLEAN COMPANY** Yourtown, MD	Date:	Price Per	PRICE
		ITEM			
				TOTAL	

3. Write a bill for a call for parts for a low-head hydro system. The delivery charge is $45.00. Gaskets cost $18.45 each. 3 gaskets were ordered. PVC piping (6 inch) was ordered. 75 feet of piping cost $2.25 per foot. PVC sealant runs $12.95 per can. 5 cans were ordered.

_____

Qty	ITEM	Price Per	PRICE
No.	**WE HAVE IT SUPPLY COMPANY** Ourtown. U.S.A.	Date:	
		TOTAL	

4. Write a bill for the 5-J Construction Company (a preferred customer) who will be given an 8% discount. Breaker boxes cost $62.50, 250 ft of 10/3 wire costing $2.73/ft, a box of screws at $15.95, and 12 electrical connectors at $0.49 each. There will be a 6% sales tax on the total bill (after the discount).

No.	KEEP IT GREEN AND CLEAN COMPANY Yourtown, MD		Date:	
Qty	**ITEM**		**Price Per**	**PRICE**
		TOTAL		

5. A pump needs to be reworked on a geothermal system. The pickup and drop-off of the pump will cost $65.00. A new shaft will be $17.45. A can of sealant will run $14.34. Labor involved is 3.5 hours at $35.00/hr. Write the bill. _____

No.	**WE HAVE IT SUPPLY COMPANY** Ourtown. U.S.A.		Date:
Qty	**ITEM**	**Price Per**	**PRICE**
	TOTAL		

6. Write an estimate for the installation of a small wind turbine. The cost of the turbine generator is $6,200.00. A 35 ft pole is $4,525.00. The electrical equipment costs $80.00. A bolt kit costs $475.00. The concrete pad for the tower costs $945.00.

Qty	No.	KEEP IT GREEN AND CLEAN COMPANY Yourtown, MD	Date:	
Qty		**ITEM**	**Price Per**	**PRICE**
			TOTAL	

7. Prepare a bill for the repair of a PV panel. The service call costs $50.00. Labor for the removal and installation of the new panel will be 4 hours at $47.00 per hour. The new panel costs $1,480.00. Wire and connectors cost an additional $43.75. There will be a 7% sales tax on the total. _____

		WE HAVE IT SUPPLY COMPANY		Date:	
No.		Ourtown. U.S.A.			
Qty		**ITEM**		**Price Per**	**PRICE**
				TOTAL	

8. Write a bill for the replacement of a circulation pump in a solar hot water heating system. The pump will run $258.00. Repair supplies run $29.85. State tax is 8%.

| No. | WE HAVE IT SUPPLY COMPANY | | Date: | |
| | Ourtown. U.S.A. | | | |
Qty	ITEM		Price Per	PRICE
			TOTAL	

9. A house is to be backfitted for a geothermal system. The drilling of 2 wells cost $7,500.00 per well. The piping for these wells is 4 lines, each 325 feet long. The tubing for the wells costs $2.93/ft. Grout and sand for the wells requires 135 bags of grout at $27.00 per bag and 26 tons of sand at $12.75 per ton. Labor would be 60 hours at $42.00 per hour. State tax is 6% on parts. Labor and drilling are not taxed. Create the bill for this.

No.	KEEP IT GREEN AND CLEAN COMPANY Yourtown, MD	Date:	
Qty	**ITEM**	**Price Per**	**PRICE**
		TOTAL	

10. A replacement storage tank and piping is needed for a solar hot water system. A 265-gallon storage tank costs $4,332.95. 1-inch copper tubing is priced at $4.83 per foot. 14 inches of tubing are needed for the job. Write up a bill for this sale.

No.	WE HAVE IT SUPPLY COMPANY	Date:	
	Ourtown. U.S.A.		
Qty	ITEM	Price Per	PRICE
W		TOTAL	

APPENDIX

Section I: Denominate Numbers

Denominate numbers are numbers that include units of measurement. The units of measurement are arranged from the largest unit at the left to the smallest unit at the right.

EXAMPLE: 6 yd 2 ft 4 in

All basic operations of arithmetic can be performed on denominate numbers.

I. Equivalent Measures

Measurements that are equal can be expressed in different terms. For example, 12 in = 1 ft. If these equivalents are divided, the answer is 1.

$$\frac{1\ ft}{12\ in} = 1 \qquad \frac{12\ in}{1\ ft} = 1$$

To express one measurement as another equal measurement, multiply by the equivalent in the form of 1.

To express 6 inches in equivalent foot measurement, multiply 6 inches by 1 in the form of $\frac{1\ ft}{12\ in}$. In the numerator and denominator, divide by a common factor.

$$6\ in = \frac{\overset{1}{\cancel{6\ in}}}{1} \times \frac{1\ ft}{\underset{2}{\cancel{12\ in}}} = \frac{1}{2}\ ft\ or\ 0.5\ ft$$

To express 4 feet in equivalent inch measurement, multiply 4 feet by 1 in the form of $\frac{12\,in}{1\,ft}$.

$$4\ ft = \frac{\overset{4}{\cancel{4\,ft}}}{1} \times \frac{12\ in}{\underset{1}{\cancel{1\,ft}}} = \frac{48\ in}{1} = 48\ in$$

Per means division, as with a fraction bar. For example, 50 miles per hour can be written $\dfrac{50\ mi}{1\ hr}$.

II. Basic Operations

A. ADDITION

EXAMPLE: 2 yd 1 ft 5 in + 1 ft 8 in + 5 yd 2 ft

1. Write the denominate numbers in a column with like units in the same column.
2. Add the denominate numbers in each column.
3. Express the answer using the largest possible units.

	2 yd	1 ft	5 in
		1 ft	8 in
+	5 yd	2 ft	
	7 yd	4 ft	13 in

7 yd			=	7 yd		
	4 ft		=	1 yd	1 ft	
		13 in	= +		1 ft	1 in
7 yd	4 ft	13 in	=	8 yd	2 ft	1 in

B. SUBTRACTION

EXAMPLE: 4 yd 3 ft 5 in − 2 yd 1 ft 7 in

1. Write the denominate numbers in columns with like units in the same column.

	4 yd	3 ft	5 in	
−	2 yd	1 ft	7 in	7 in is larger than 5 in

2. Starting at the right, examine each column to compare the numbers. If the bottom number is larger, exchange one unit from the column at the left for its equivalent. Combine like units.

3 ft = 2 ft 12 in

12 in + 5 in = 17 in

3. Subtract the denominate numbers.

$$
\begin{array}{r}
4 \text{ yd} \quad 2 \text{ ft} \quad 17 \text{ in} \\
- \; 2 \text{ yd} \quad 1 \text{ ft} \quad 7 \text{ in} \\
\hline
2 \text{ yd} \quad 1 \text{ ft} \quad 10 \text{ in}
\end{array}
$$

4. Express the answer using the largest possible units.

2 yd   1 ft   10 in

C. MULTIPLICATION

—By a constant

EXAMPLE: 1 hr 24 min × 3

1. Multiply the denominate number by the constant.

$$
\begin{array}{r}
1 \text{ hr} \quad 24 \text{ min} \\
\times \; 3 \\
\hline
3 \text{ hr} \quad 72 \text{ min}
\end{array}
$$

2. Express the answer using the largest possible units.

$$
\begin{array}{rcll}
3 \text{ hr} & & = & 3 \text{ hr} \\
72 \text{ min} & & = & 1 \text{ hr} \quad 12 \text{ min} \\
\hline
3 \text{ hr} \quad 72 \text{ min} & & = & 4 \text{ hr} \quad 12 \text{ min}
\end{array}
$$

—By a denominate number expressing linear measurement

EXAMPLE: 9 ft 6 in × 10 ft

1. Express all denominate numbers in the same unit.

$$9 \text{ ft } 6 \text{ in} = 9\frac{1}{2} \text{ ft}$$

2. Multiply the denominate numbers. (This includes the units of measure, such as ft × ft = sq ft.)

$$9\frac{1}{2} \text{ ft} \times 10 \text{ ft} =$$

$$\frac{19}{2} \text{ ft} \times 10 \text{ ft} =$$

95 sq ft

—By a denominate number expressing square measurement

EXAMPLE: 3 ft × 6 sq ft

1. Multiply the denominate numbers.   3 ft × 6 sq ft = 18 cu ft
   (This includes the units of measure,
   such as ft × ft = sq ft and
   sq ft × ft = cu ft.)

—By a denominate number expressing rate

EXAMPLE: 50 mi per hr × 3 hours

1. Express the rate as a fraction using $\dfrac{50 \text{ mi}}{1 \text{ hr}} \times \dfrac{3 \text{ hr}}{1}$
   the fraction bar for *per.*

2. Divide the numerator and denominator $\dfrac{50 \text{ mi}}{\cancel{1 \text{ hr}}_{1}} \times \dfrac{\cancel{3 \text{ hr}}^{3}}{1}$
   by any common factors, including
   units of measure.

3. Multiply numerators. $\dfrac{150 \text{ mi}}{1} =$
   Multiply denominators.

4. Express the answer in the remaining unit.   150 mi

D. DIVISION

—By a constant

EXAMPLE: 8 gal 3 qt ÷ 5

1. Express all denominate numbers   8 gal 3 qt = 35 qt
   in the same unit.
2. Divide the denominate number   35 qt ÷ 5 = 7 qt
   by the constant.
3. Express the answer using the   7 qt = 1 gal 3 qt
   largest possible units.

—By a denominate number expressing linear measurement

EXAMPLE: 11 ft 4 in ÷ 8 in

1. Express all denominate numbers
   in the same unit.

   11 ft 4 in = 136 in

2. Divide the denominate numbers
   by a common factor. (This includes
   the units of measure, such as inches
   ÷ inches = 1.)

   136 in ÷ 8 in =

   $$\dfrac{\overset{17}{\cancel{136}\,\cancel{\text{in}}}}{\underset{1}{\cancel{8}\,\cancel{\text{in}}}} = \dfrac{17}{1} = 17$$

—By a linear measure with a square measurement as the dividend

EXAMPLE: 20 sq ft ÷ 4 ft

1. Divide the denominate numbers.
   (This includes the units of measure,
   such as sq ft ÷ ft = ft.)

   20 sq ft ÷ 4 ft

   $$\dfrac{\overset{5\ \text{ft}}{\cancel{20}\,\cancel{\text{sq ft}}}}{\cancel{4}\,\cancel{\text{ft}}} = \dfrac{5\ \text{ft}}{1}$$

2. Express the answer in the remaining unit. 5 ft

—By denominate numbers used to find rate

EXAMPLE: 200 mi ÷ 10 gal

1. Divide the denominate numbers.

   $$\dfrac{\overset{20\ \text{mi}}{\cancel{200}\,\cancel{\text{mi}}}}{\underset{1\ \text{gal}}{\cancel{10}\,\cancel{\text{gal}}}} = \dfrac{20\ \text{mi}}{1\ \text{gal}}$$

2. Express the units with the fraction
   bar meaning *per*.

   $$\dfrac{20\ \text{mi}}{1\ \text{gal}} = 20\ \text{miles per gallon}$$

NOTE: Alternate methods of performing the basic operations will produce the same results. The choice of method is determined by the individual.

Section II: Equivalent Values

Equivalent English Relationships

LENGTH EQUIVALENTS

$\frac{1}{16}$ inch	=	0.0625	inch
$\frac{1}{8}$ inch	=	0.125	inch
$\frac{3}{16}$ inch	=	0.1875	inch
$\frac{1}{4}$ inch	=	0.25	inch
$\frac{5}{16}$ inch	=	0.3125	inch
$\frac{3}{8}$ inch	=	0.375	inch
$\frac{7}{16}$ inch	=	0.4375	inch
$\frac{1}{2}$ inch	=	0.5	inch
$\frac{9}{16}$ inch	=	0.5625	inch
$\frac{5}{8}$ inch	=	0.625	inch
$\frac{11}{16}$ inch	=	0.6875	inch
$\frac{3}{4}$ inch	=	0.75	inch
$\frac{13}{16}$ inch	=	0.8125	inch
$\frac{7}{8}$ inch	=	0.875	inch
$\frac{15}{16}$ inch	=	0.9375	inch
1 inch	=	0.0833	foot
2 inches	=	0.1666	foot
3 inches	=	0.25	foot
4 inches	=	0.3333	foot
5 inches	=	0.4166	foot
6 inches	=	0.5	foot
7 inches	=	0.5833	foot
8 inches	=	0.6666	foot
9 inches	=	0.75	foot
10 inches	=	0.8333	foot
11 inches	=	0.9166	foot

ENGLISH LENGTH MEASURE

1 foot (ft)	=	12 inches (in)
1 yard (yd)	=	3 feet (ft)
1 mile (mi)	=	1,760 yards (yd)
1 mile (mi)	=	5,280 feet (ft)

ENGLISH AREA MEASURE

1 square yard (sq yd)	=	9 square feet (sq ft)
1 square foot (sq ft)	=	144 square inches (sq in)
1 square mile (sq mi)	=	640 acres
1 acre	=	43,560 square feet (sq ft)

ENGLISH VOLUME MEASURE FOR SOLIDS

1 cubic yard (cu yd)	=	27 cubic feet (cu ft)
1 cubic foot (cu ft)	=	1,728 cubic inches (cu in)

ENGLISH VOLUME MEASURE FOR FLUIDS

1 quart (qt)	=	2 pints (pt)
1 gallon (gal)	=	4 quarts (qt)

ENGLISH VOLUME MEASURE EQUIVALENTS

1 gallon (gal)	=	0.133681 cubic foot (cu ft)
1 gallon (gal)	=	231 cubic inches (cu in)

ENGLISH WEIGHT (MASS) MEASURE EQUIVALENTS

1 pound (lb)	=	16 ounces (oz)

SI Metrics Style Guide

SI metrics is derived from the French name Le Système International d'Unités. The metric unit names are already in accepted practice. SI metrics attempts to standardize the names and usages so that students of metrics will have a universal knowledge of the application of terms, symbols, and units.

The English system of mathematics (used in the United States) has always had many units in its weights and measures tables that were not applied to everyday use. For example, the pole, perch, furlong, peck, and scruple are not used often. These measurements, however, are used to form other measurements, and it has been necessary to include the measurements in the tables. Including these measurements aids in the understanding of the orderly sequence of measurements greater or smaller than the less frequently used units.

The metric system also has units that are not used in everyday application. Only by learning the lesser-used units is it possible to understand the order of the metric system. SI metrics, however, places an emphasis on the most frequently used units.

In using the metric system and writing its symbols, certain guidelines are followed. For the student's reference, some of the guidelines are listed.

1. In using the symbols for metric units, the first letter is capitalized only if it is derived from the name of a person.

EXAMPLE:

UNIT	SYMBOL	UNIT	SYMBOL
meter	m	Newton	N (named after Sir Isaac Newton)
gram	g	degree Celsius	°C (named after Anders Celsius)

EXCEPTIONS: The symbol for liter is L. This is used to distinguish it from the number one (1).

2. Prefixes are written with lowercase letters.

EXAMPLE:

PREFIX	UNIT	SYMBOL
centi	meter	cm
milli	gram	mg

EXCEPTIONS:

PREFIX	UNIT	SYMBOL
tera	meter	Tm (used to distinguish it from the metric tonne, t)
giga	meter	Gm (used to distinguish it from gram, g)
mega	gram	Mg (used to distinguish it from milli, m)

3. Periods are not used in the symbols. Symbols for units are the same in the singular and the plural (no "s" is added to indicate a plural).

EXAMPLE: 1 mm   *not*   1 mm.
               3 mm   *not*   3 mms

4. When referring to a unit of measurement, symbols are not used. The symbol is used only when a number is associated with it.

EXAMPLE: The length of the room is   *not*   The length of the room is expressed in m.
               expressed in meters.   (*The length of the room is 25 m* is correct.)

5. When writing measurements that are less than 1, a zero is written before the decimal point.

EXAMPLE: 0.25 m   *not*   .25 m

6. A space is left between the digits and the unit of measure.

EXAMPLE: 5,179,232 mm   *not*   5,179,232mm

7. Symbols for area measure and volume measure are written with exponents.

EXAMPLE: 3 cm^3 *not*   3 sq cm
               4 km^3 *not*   4 cu km

8. Metric words with prefixes are accented on the first syllable. In particular, kilometer is pronounced "kill'-o-meter." This avoids confusion with words for measuring devices, which are generally accented on the second syllable, such as thermometer (ther-mom'-e-ter).

Metric Relationships

The base units in SI metrics include the meter and the gram. Other units of measure are related to these units. The relationship between the units is based on powers of ten and uses these prefixes:

kilo (1,000)   centi (0.01)   milli (0.001)

These tables show the most frequently used units with an asterisk (*).

METRIC LENGTH MEASURE

10 millimeters (mm)*	= 1 centimeter (cm)*
100 centimeters (cm)	= 1 meter (m)*
1,000 meters (m)	= 1 kilometer (km)*

- To express a metric length unit as a smaller metric length unit, multiply by a positive power of ten such as 10; 100; 1,000; 10,000; and so on.

- To express a metric length unit as a larger metric length unit, multiply by a negative power of ten such as 0.1; 0.01; 0.001; 0.0001; and so on.

METRIC AREA MEASURE

100 square millimeters (mm^2)	= 1 square centimeter (cm^2)
10,000 square centimeters (cm^2)	= 1 square meter (m^2)
10,000 square meters (m^2)	= 1 square kilometer (km^2)

- To express a metric area unit as a smaller metric area unit, multiply by 100; 10,000; 1,000,000; and so on.

- To express a metric area unit as a larger metric area unit, multiply by 0.01; 0.0001; 0.000001; and so on.

METRIC VOLUME MEASURE FOR SOLIDS

1,000 cubic millimeters (mm^3)	= 1 cubic centimeter (cm^3)*
1,000,000 cubic centimeters (cm^3)	= 1 cubic meter (m^3)*
1,000,000,000 cubic meters (m^3)	= 1 cubic kilometer (km^3)

- To express a metric volume unit for solids as a smaller metric volume unit for solids, multiply by 1,000; 1,000,000; 1,000,000,000; and so on.

- To express a metric volume unit for solids as a larger metric volume unit for solids, multiply by 0.001; 0.000001; 0.000000001; and so on.

METRIC VOLUME MEASURE FOR FLUIDS

100 milliliters (mL)*	= 1 centiliter (cL)
100 centiliters (cL)	= 1 liter (L)*
1,000 liters (L)	= 1 kiloliter (kL)

- To express a metric volume unit for fluids as a smaller metric volume unit for fluids, multiply by 10; 100; 1,000; 10,000; and so on.

- To express a metric volume unit for fluids as a larger metric volume unit for fluids, multiply by 0.1; 0.01; 0.001; 0.0001; and so on.

METRIC VOLUME MEASURE EQUIVALENTS

1,000 cubic centimeters (cm^3)	= 1 liter (L)
1 cubic centimeter (cm^3)	= 1 milliliter (mL)

METRIC MASS MEASURE

10 milligrams (mg)*	= 1 centigram (cg)
100 centigrams (cg)	= 1 gram (g)*
1,000 grams (g)	= 1 kilogram (kg)*
1,000 kilograms (kg)	= 1 megagram (Mg)*

- To express a metric mass unit as a smaller metric mass unit, multiply by 10; 100; 1,000; 10,000; and so on.

- To express a metric mass unit as a larger metric mass unit, multiply by 0.1; 0.01; 0.001; 0.0001; and so on.

Metric measurements are expressed in decimal parts of a whole number. For example, one-half millimeter is written as 0.5 mm.

In calculating with the metric system, all measurements are expressed using the same prefixes. If answers are needed in millimeters, all parts of the problem should be expressed in millimeters before the final solution is attempted. Diagrams that have dimensions in different prefixes must first be expressed using the same unit.

English–Metric Equivalents

Length Measure

1 inch (in)	=	25.4 millimeters (mm)
1 inch (in)	=	2.54 centimeters (cm)
1 inch (in)	=	0.0254 meter (m)
1 foot (ft)	=	0.3048 meter (m)
1 yard (yd)	=	0.9144 meter (m)
1 mile (mi)	=	1.609 kilometers (km)
1 millimeter (mm)	=	0.03937 inch (in)
1 centimeter (cm)	=	0.3937 inch (in)
1 meter (m)	=	39.37008 inches (in)
1 meter (m)	=	3.28084 feet (ft)
1 meter (m)	=	1.09361 yards (yd)
1 kilometer (km)	=	0.62137 mile (mi)

Area Measure

1 square inch (sq in)	=	645.16 square millimeters (mm^2)
1 square inch (sq in)	=	6.4516 square centimeters (cm^2)
1 square foot (sq ft)	=	0.092903 square meter (m^2)
1 square yard (sq yd)	=	0.836127 square meter (m^2)
1 square millimeter (mm^2)	=	0.00155 square inch (sq in)
1 square centimeter (cm^2)	=	0.155 square inch (sq in)
1 square meter (m^2)	=	10.76391 square feet (sq ft)
1 square meter (m^2)	=	1.19599 square yards (sq yd)

Volume Measure For Solids

1 cubic inch (cu in)	=	16.387064 cubic centimeters (cm^3)
1 cubic foot (cu ft)	=	0.028317 cubic meter (m^3)
1 cubic foot (cu ft)	=	28,317 cubic centimeters (cm^3)
1 cubic yard (cu yd)	=	0.764555 cubic meter (m^3)
1 cubic centimeter (cm^3)	=	0.061024 cubic inch (cu in)
1 cubic meter (m^3)	=	61,023 cubic inches (cu in)
1 cubic meter (m^3)	=	35.314667 cubic feet (cu ft)
1 cubic meter (m^3)	=	1.307951 cubic yards (cu yd)

Volume Measure For Fluids

1 gallon (gal)	=	3,785.411 cubic centimeters (cm^3)
1 gallon (gal)	=	3.785411 liters (L)
1 quart (qt)	=	0.946353 liter (L)
1 ounce (oz)	=	29.57353 cubic centimeters (cm^3)
1 cubic centimeter (cm^3)	=	0.000264 gallon (gal)
1 liter (L)	=	0.264172 gallon (gal)
1 liter (L)	=	1.056688 quarts (qt)
1 cubic centimeter (cm^3)	=	0.033814 ounce (oz)

Mass Measure

1 pound (lb)	=	0.453592 kilogram (kg)
1 pound (lb)	=	453.59237 grams (g)
1 ounce (oz)	=	28.349523 grams (g)
1 ounce (oz)	=	0.02835 kilogram (kg)
1 kilogram (kg)	=	2.204623 pounds (lb)
1 gram (g)	=	0.002205 pound (lb)
1 kilogram (kg)	=	35.273962 ounces (oz)
1 gram (g)	=	0.035274 ounce (oz)

Power Measure

1 horsepower	=	746 watts

Section III: Instruments for Measuring

Frequently, lengths or distances must be measured. You need to measure accurately. Two devices used for this purpose are the ruler and the tape. Being able to read them is a necessity.

Measuring with a Ruler or Tape

Rulers tend to be short in length—as short as 6 inches to as long as a little over 39 inches. Most rulers have two scales marked on them. One is the metric scale, and the other is the English scale.

The metric scale can be identified by the marking of cm or centimeters near the scale. If no markings are shown, the metric scales have the higher numbers printed on it. The metric scale is based on groups of 10, so there are longer marks on the scale with numbers alongside the marks. These are centimeters. There are 9 shorter marks in between the long marks. Very often, the middle mark will be a little bit longer than the other 8 marks. Each of these shorter marks is a tenth of a centimeter or a millimeter. The length is given as a decimal number. Count the number of marks past the cm line. This is the decimal part of the measurement.

EXAMPLE: Determine the reading on the ruler below.

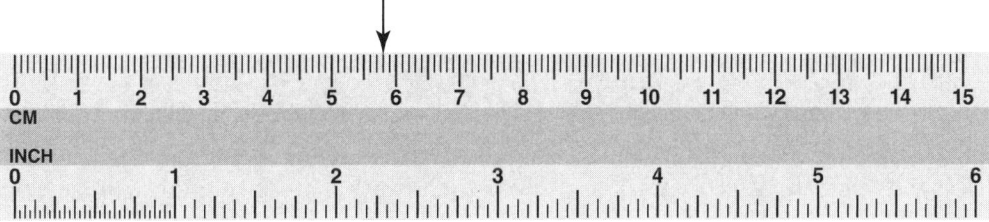

The arrow points to a mark that is between the 5 and the 6, so it is greater than 5 cm but not 6 cm. The mark is the 8th mark past the 5, making the reading 5.8 cm.

The English scale is often marked inches. Reading the inch scale is a little more involved. The marks are a series of different lengths. They are based on fractions (powers) of $\frac{1}{2}$. The longest marks are the inch marks and have numbers next to them. The next longest mark is located half-way between the inch marks and is the half-inch mark. Then there are 2 marks a little shorter in length. The first is found half way between the inch and the half-inch mark. It is half of the half inch or a quarter-inch mark ($\frac{1}{2} \times \frac{1}{2} = \frac{1}{4}$). There is another quarter-inch mark located half way between the half-inch mark and the next inch mark. This mark is the $\frac{3}{4}$ inch mark. These marks are the $\frac{1}{4}$ and $\frac{3}{4}$ inch mark. The $\frac{2}{4}$ inch mark is the $\frac{1}{2}$ inch mark.

Half way between the inch mark and the $\frac{1}{4}$ inch mark is a shorter mark. It is ($\frac{1}{2} \times \frac{1}{4} = \frac{1}{8}$) the eighth-inch mark. There are the same size marks at the $\frac{3}{8}$, $\frac{5}{8}$, and $\frac{7}{8}$ inch positions (the $\frac{2}{8}$, $\frac{4}{8}$, and $\frac{6}{8}$ marks reduce to $\frac{1}{4}$, $\frac{1}{2}$, and $\frac{3}{4}$, respectively, and are the larger marks). Finally, most rulers have the shortest mark halfway between the inch mark and the $\frac{1}{8}$-inch mark. This is the ($\frac{1}{2} \times \frac{1}{8} = \frac{1}{16}$) $\frac{1}{16}$-inch mark. Similar to what we have been saying, the shortest marks are the $\frac{1}{16}$, $\frac{3}{16}$, $\frac{5}{16}$, $\frac{7}{16}$, $\frac{9}{16}$, $\frac{11}{16}$, $\frac{13}{16}$, and $\frac{15}{16}$ inch marks. All of the rest reduce to marks we have already talked about.

When making a measurement, align the beginning of the ruler with one end of the distance to be measured. Find the mark that lines up with the other end. Determine what type of mark (inch, $\frac{1}{2}$ inch, $\frac{1}{4}$ inch, $\frac{1}{8}$ inch, $\frac{1}{16}$ inch) it is and what number it is from the inch mark. You also need to determine which inch mark (if any) you have gone past. It is not the closest inch mark that is used for the measurement but the highest inch mark that has been passed. (Always line up the beginning of the ruler first.)

EXAMPLE: Determine the reading indicated by the arrow.

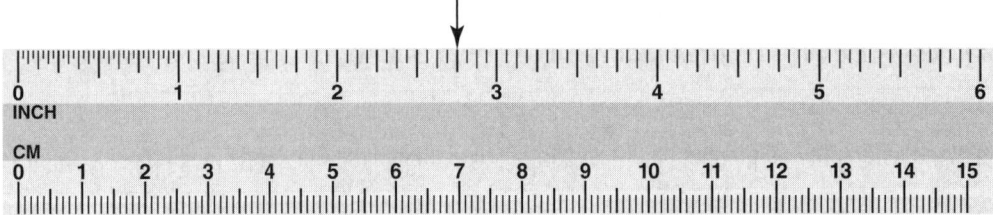

The mark is between the 2-inch mark and the 3-inch mark, so the reading will be 2 plus some fraction inches. The longest mark in between the inch marks is the $\frac{1}{2}$ inch mark. This is the next shorter mark, and it is between the $\frac{1}{2}$ mark and the 3-inch mark. The mark is a $\frac{1}{4}$ inch mark, but because it is greater than $\frac{1}{2}$, it will be the $\frac{3}{4}$ mark. So, the reading is $2\frac{3}{4}$ inches.

EXAMPLE: Determine the reading indicated by the arrow.

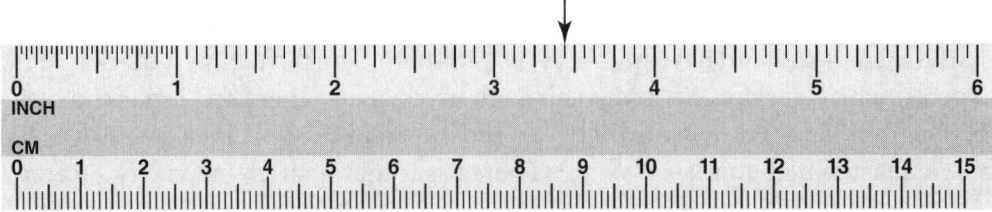

This mark is between 3 and 4 inches, so the reading will be 3 and some fraction inches. The mark is the shortest mark, so in our case, it will be a 16th inch reading. Counting each from the 3-inch mark, we count 7 marks, making the fraction $\frac{7}{16}$. Our reading is $3\frac{7}{16}$ inches.

For measuring longer distances, a tape is often used. The tape is usually metal and is many feet long. Examples of tape lengths are 10', 12', 25', 50', and 100'. Often with these tapes, there are two sets of markings. The first is just a continuous increase in the inch numbering. The second is an increase in the inch marking until 11 inches. The next mark is shown as 1 foot. The next marking is 1 inch, then 2, then 3, and so on until 11. The next mark is 2 feet. Then it starts with 1 again. This scale is giving the feet and inches. Measuring with a tape is similar to measuring with a ruler with the exception that one can include the foot length as well as the inch length. Reading the foot length is done in the same manner as the inches, that is, the value given is the last one passed, not necessarily the closest one to where you are measuring.

EXAMPLE: Determine the reading indicated by the arrow.

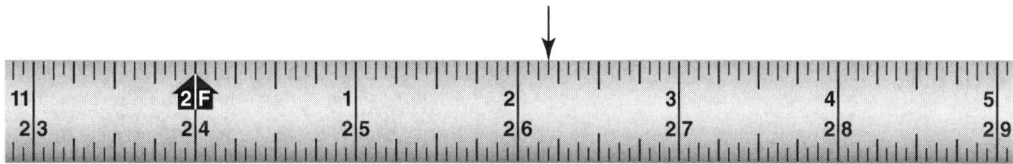

The mark is past the 2 ft indication and past the 2 in indication. It is the third mark, and it is one of the shortest marks, so it is the $\frac{3}{16}$ mark. This makes the reading 2 ft $2\frac{3}{16}$ in. It could just as easily have been reported as $26\frac{3}{16}$ in.

At times, even more precise measurements are needed than we can achieve with a measuring tape. Two devices used to get such precise measurements are the vernier caliper and the micrometer. The challenge is to read scales. So, let us make sure that we can use and read the scales.

Measuring with a Vernier Caliper

To read the scale on the vernier caliper, look first at where the 0 line is lining up on the main scale. This gives the first two numbers. Then look along the vernier scale to find the line that most closely lines up with one of the lines on the main scale. This will give the next two digits.

EXAMPLE: Determine the reading on the vernier caliper in the photo. The reading is needed in inches, so the lower two scales are used. (The upper two scales are used for millimeter readings.)

A.   The 0 line on the vernier scale is just past the 7, which is *before* the large 1 on the main scale. Because it is before the large 1, the reading begins 0.70.

B.   The 0 line is *before* the short mark indicating 7.5, so the next part of the reading will be less than 50.

C.   Looking down the vernier scale, we see that the 29 line on the vernier exactly matches a line on the main scale (in this case, the 2.15 line). At the bottom of the vernier is an indication .001 in. This means that each mark on the vernier scale represents 0.001 inch. Because the 29 line matches, this represents 0.029 in.

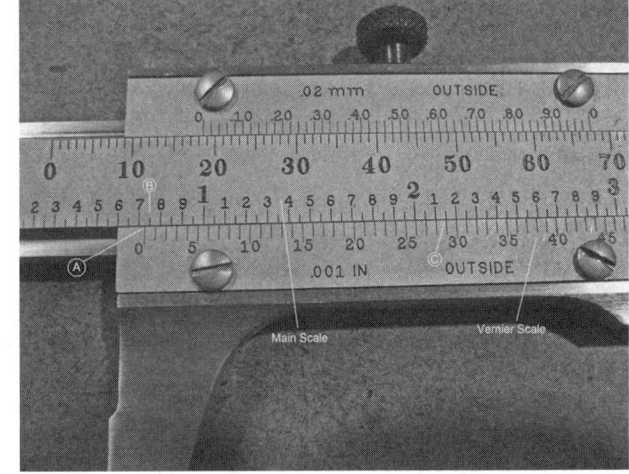

D.   Combine the readings from step A and step C.

$$0.7 + 0.029 = 0.729$$

The final reading is 0.729 inch.

NOTE: If the 0 line had been past the little mark in part B, the reading would have been 0.75 plus 0.029 to give a final reading of 0.779 inch.

Measuring with a Micrometer

The micrometer is another device used to make accurate measurements. To read the scale, the first part of the reading is determined by what is uncovered by the rotating barrel as it unwinds. The rest of the reading is determined by where the main scale line lines up with the markings on the barrel.

The numbers on the main (stationary) part of the scale give the first digit and part of the rest of the reading. The printed numbers on the main scale of this micrometer are tenths of an inch. Notice that the distance between two numbers is divided into 4 quarters. Each mark (division) on the main scale between the tenths marks represents 0.025 inch. The rotating barrel of the micrometer has 25 equal divisions on it. One complete revolution uncovers one division on the main scale. One revolution covers 0.025 inch, so each mark on the barrel represents 0.001 inch.

EXAMPLE: Determine the reading on the micrometer in the figure below.

A. The barrel (rotating part) has uncovered the 2. This represents 0.2.
B. The barrel has also uncovered two divisions past the 2. This means that the reading is going to be 2 × 0.025 or 0.05 more than the 0.2.
C. The main scale line lines up with the 6 on the rotating barrel. This gives an additional 6 × 0.001 or 0.006 inch.
D. Combine these three parts to give the total reading.

0.2 + 0.05 + 0.006 = 0.256 inch

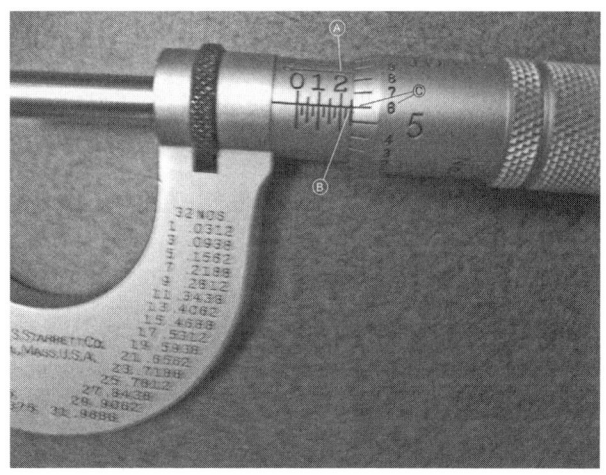

If the rotating barrel in our picture had uncovered 3 divisions after the 2, the reading would be 0.2 + 0.075 + 0.006 or 0.281 inch. If the rotating barrel in this picture had shown a number such as 14, then the reading would be 0.2 + 0.05 + 0.014 giving a total of 0.264 inches. As can be seen, the micrometer has to be read carefully. Some micrometers have lines on the back side of the main part of the scale. This set of lines is a vernier that will add an additional digit if the marks on the barrel do not line up exactly with the main scale line. The additional digit is determined by a line on the vernier lining up with a line on the rotating barrel.

Section IV: Formulas

Perimeter

Square

$P = 4s$

P = perimeter

s = side

Rectangle

$P = 2l + 2w$

P = perimeter

l = length

w = width

Triangle

$P = a + b + c$

P = perimeter

a = first side

b = second side

c = third side

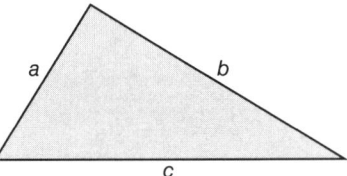

Circle

$C = 2\pi r$

$C = \pi D$

$r = \dfrac{D}{2}$

C = circumference

π = 3.1416

r = radius

D = diameter

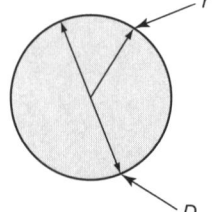

266

Area

Square
$A = s^2$

A = area
s = side

Rectangle
$A = lw$

A = area
l = length
w = width

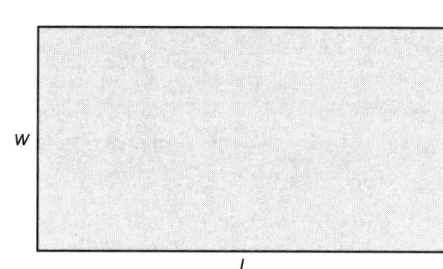

Triangle
$A = \frac{1}{2}bh$

A = area
b = base
h = height

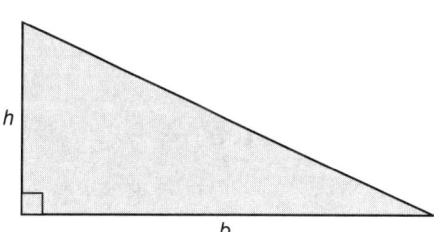

Circle
$A = \pi r^2$

A = area
π = 3.1416
r = radius
D = diameter

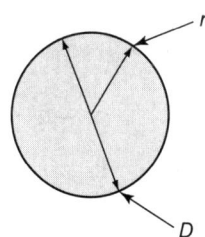

Volume

Cubic Solid
$V = s^3$

V = volume
s = side

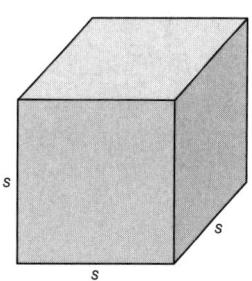

Rectangular Solid

$V = lwh$

V = volume
l = length
w = width
h = height

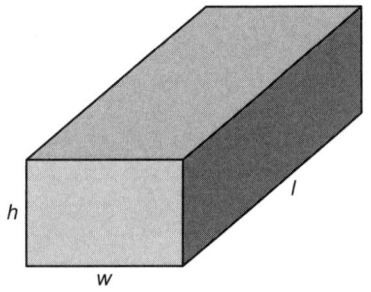

Cylindrical Solid

$V = \pi r^2 h$

V = volume

$\pi = 3.1416$
r = radius
D = diameter
h = height

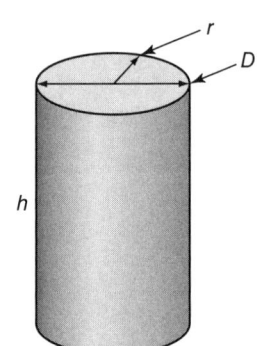

Temperature

$^\circ C = 5/9\ (^\circ F - 32)$ $^\circ C$ = degrees Celsius
$^\circ F = 9/5\ (^\circ C) + 32$ $^\circ F$ = degrees Fahrenheit

Electrical

Ohm's Law

$E = IR$

E = voltage (in volts)
I = current (in amperes)
R = resistance (in ohms)

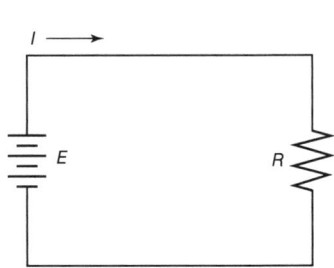

Power Formula

$P = IE$

P = power (in watts)
I = current (in amperes)
E = voltage (in volts)

Resistance in Series
$R = R_1 + R_2 + \ldots$ R = resistance (in ohms)

Resistance in Parallel

$$R = \cfrac{1}{\cfrac{1}{R_1} + \cfrac{1}{R_2} + \ldots}$$ R = resistance (in ohms)

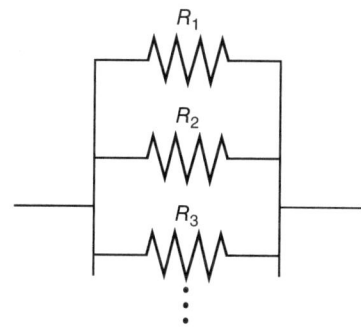

Section V: Trigonometric Functions

TRIGONOMETRIC FUNCTIONS

Angle	Sine	Cosine	Tangent	Angle	Sine	Cosine	Tangent
1°	0.0175	0.9998	0.0175	46°	0.7193	0.6947	1.0355
2°	0.0349	0.9994	0.0349	47°	0.7314	0.6820	1.0724
3°	0.0523	0.9986	0.0524	48°	0.7431	0.6691	1.1106
4°	0.0698	0.9976	0.0699	49°	0.7547	0.6561	1.1504
5°	0.0872	0.9962	0.0875	50°	0.7660	0.6428	1.1918
6°	0.1045	0.9945	0.1051	51°	0.7771	0.6293	1.2349
7°	0.1219	0.9925	0.1228	52°	0.7880	0.6157	1.2799
8°	0.1392	0.9903	0.1405	53°	0.7986	0.6018	1.3270
9°	0.1564	0.9877	0.1584	54°	0.8090	0.5878	1.3764
10°	0.1736	0.9848	0.1763	55°	0.8192	0.5736	1.4281
11°	0.1908	0.9816	0.1944	56°	0.8290	0.5592	1.4826
12°	0.2079	0.9781	0.2126	57°	0.8387	0.5446	1.5399
13°	0.2250	0.9744	0.2309	58°	0.8480	0.5299	1.6003
14°	0.2419	0.9703	0.2493	59°	0.8572	0.5150	1.6643
15°	0.2588	0.9659	0.2679	60°	0.8660	0.5000	1.7321
16°	0.2756	0.9613	0.2867	61°	0.8746	0.4848	1.8040
17°	0.2924	0.9563	0.3057	62°	0.8829	0.4695	1.8807
18°	0.3090	0.9511	0.3249	63°	0.8910	0.4540	1.9626
19°	0.3256	0.9455	0.3443	64°	0.8988	0.4384	2.0503
20°	0.3420	0.9397	0.3640	65°	0.9063	0.4226	2.1445
21°	0.3584	0.9336	0.3839	66°	0.9135	0.4067	2.2460
22°	0.3746	0.9272	0.4040	67°	0.9205	0.3907	2.3559
23°	0.3907	0.9205	0.4245	68°	0.9272	0.3746	2.4751
24°	0.4067	0.9135	0.4452	69°	0.9336	0.3584	2.6051
25°	0.4226	0.9063	0.4663	70°	0.9397	0.3420	2.7475
26°	0.4384	0.8988	0.4877	71°	0.9455	0.3256	2.9042
27°	0.4540	0.8910	0.5095	72°	0.9511	0.3090	3.0777
28°	0.4695	0.8829	0.5317	73°	0.9563	0.2924	3.2709
29°	0.4848	0.8746	0.5543	74°	0.9613	0.2756	3.4874
30°	0.5000	0.8660	0.5774	75°	0.9659	0.2588	3.7321
31°	0.5150	0.8572	0.6009	76°	0.9703	0.2419	4.0108
32°	0.5299	0.8480	0.6249	77°	0.9744	0.2250	4.3315
33°	0.5446	0.8387	0.6494	78°	0.9781	0.2079	4.7046
34°	0.5592	0.8290	0.6745	79°	0.9816	0.1908	5.1446
35°	0.5736	0.8192	0.7002	80°	0.9848	0.1736	5.6713
36°	0.5878	0.8090	0.7265	81°	0.9877	0.1564	6.3138
37°	0.6018	0.7986	0.7536	82°	0.9903	0.1392	7.1154
38°	0.6157	0.7880	0.7813	83°	0.9925	0.1219	8.1443
39°	0.6293	0.7771	0.8098	84°	0.9945	0.1045	9.5144
40°	0.6428	0.7660	0.8391	85°	0.9962	0.0872	11.4301
41°	0.6561	0.7547	0.8693	86°	0.9976	0.0698	14.3007
42°	0.6691	0.7431	0.9004	87°	0.9986	0.0523	19.0811
43°	0.6820	0.7314	0.9325	88°	0.9994	0.0349	28.6363
44°	0.6947	0.7193	0.9657	89°	0.9998	0.0175	57.2900
45°	0.7071	0.7071	1.0000	90°	1.0000	0.0000	

GLOSSARY

Ampere The unit of electric current. This is abbreviated by amp.

Anemometer A device used for measuring wind speed.

Bending radius The radius of the circle formed by the arc made when tubing is bent.

British thermal unit (Btu) A unit for a quantity of heat. It is the quantity of heat needed to raise the temperature of 1 pound of water 1 degree Fahrenheit.

Calorie A unit for the quantity of heat. It is the quantity of heat needed to raise the temperature of 1 gram of water 1 degree Celsius.

Circulation pump A pump used to push a liquid through one or more elements, with the path returning to the pump.

Coefficient of performance A measure of the efficiency of a heat pump found by comparing the amount of heat that is moved (into the building during heating or out of the building during cooling) and the energy used by the pumping system.

Conduit A tube used to carry and protect electrical wires.

Dam breast The structure built to hold back water from flowing in a stream, creating a pool.

ΔT The change in temperature. This is found by subtracting the final temperature from the initial temperature.

Elbow A section of pipe that bends at some angle.

Foot-pound A unit of energy or work. This is equal to the work done by one pound of weight lowering one foot.

Geothermal heating system A heating system where the source of heat is the earth.

Gpm a unit used as a measure of fluid flow. The abbreviation is for gallons per minute.

Ground source heat pump A heat pump that takes heat from or exhausts heat to the earth.

Grout A kind of cement used to fill a well and improve the heat transfer properties with the earth.

Guy wire A wire between a pole and an anchor used to help support the pole.

HDPE High density polyethylene—a plastic material used to make pipes or tubes.

Heat exchange The movement of heat from one material to another.

271

Horizontal ground loop A loop of tubing used to exchange heat between a fluid and the earth buried at a constant depth below the surface.

Joule A metric unit of energy or work. It is the name given to the quantity of 1 kilogram meter squared per second squared.

kW Kilowatt—one thousand watts—a larger unit of power.

kWh Kilowatt hour—a unit of energy. The product of power times time.

Low-head hydro turbine A turbine specifically designed to work using water falling from a height of about 10–20 meters or less (33–66 feet).

MWh The abbreviation for megawatt hour. This is one thousand kWh.

Ohm A unit of electrical resistance. Or the unit of resistance to electrical flow.

Open loop geothermal system A system that puts water into a well and takes water out of a different place in the well for heating or cooling.

Penstock The piping from a dam or water source to the turbine of the generator.

Photovoltaic (PV) A device or system that changes light energy directly into electrical energy.

Power The rate at which work is done. The British units are foot-pounds per second. A second unit is horsepower. The metric unit is watt.

PVC Polyvinyl chloride—a plastic used to make pipes and tubes.

Slug An English unit of mass.

Slurry A thin, watery mixture such as cement before it sets and hardens.

Solar collector A device used to capture solar energy and convert it either to heat energy or to electrical energy.

Solar insolation The amount of sunlight energy falling on each square unit of surface in a unit time.

Turbine footprint The space on the ground that is needed for a wind turbine to operate properly with no interference from another turbine.

U Factor A measure of the ease for heat to flow through a barrier.

Volt An unit of electrical potential or electromotive force.

Watt A metric unit of power. This is the rate at which work is done. A watt is another name for a kilogram meter squared per second cubed.

Wind farm A location where multiple wind turbines have been located.

Wind turbine A device that changes wind energy to rotational (turning) energy. It can be used to turn a generator, creating electrical energy.

ANSWERS
TO ODD-NUMBERED PROBLEMS

Section 1
Whole Numbers

Unit 1
Addition of Whole Numbers

1. 938
3. 11,005
5. 10,081 feet
7. 1,232
9. 6,998
11. 7,892 miles
13. 1,071 ft
15. 53 hours
17. 84 ft
19. 168 miles
21. 557 gallons
23. 500 ft
25. 827 ft

Unit 2
Subtraction of Whole Numbers

1. 436
3. 6,248
5. 2,545 MW
7. 214
9. 4,489
11. 5,831 feet
13. 68 feet
15. 515 feet

17. 692 kWh
19. 47 ft
21. 72 square feet
23. 126 ft
25. 615 feet

Unit 3
Multiplication of Whole Numbers

1. 416
3. 190,099
5. 192 square feet
7. 7,084
9. 230,622
11. 2,236 square centimeters
13. 360,000 pounds
15. 3,096 feet
17. 476 hours
19. 2,540 ft
21. 90,720 pounds
23. 3,024,000 gallons
25. 37,800 kWh

Unit 4
Division of Whole Numbers

1. 213
3. 335

5. 410 feet
7. 82
9. 463
11. 438 sq in
13. 17 poles
15. 7 gallons per second
17. 23 turbines
19. 16 hours
21. A. 8 sides
     B. 20 feet
23. 175 feet per ton
25. 180 gallons in one branch

Unit 5
Combined Operations with Whole Numbers

1. 1,722
3. 5,314
5. 32,524
7. 367
9. 3,923
11. 1,119,528
13. 8,122 MW
15. 876 ft
17. 58 ft
19. 61,560 W
21. 630 sq ft
23. 60 cu ft
25. 14 days

Section 2
Common Fractions

Unit 6
Addition of Common Fractions

1. $\frac{7}{9}$
3. $3\frac{1}{2}$
5. $1\frac{1}{4}$ hours
7. $1\frac{2}{7}$
9. $5\frac{17}{22}$
11. $1\frac{1}{2}$ meters
13. $\frac{5}{6}$
15. $\frac{7}{12}$
17. $124\frac{3}{8}$ gallons
19. $28\frac{2}{3}$ ft
21. $66\frac{1}{16}''$
23. $\frac{23}{24}$
25. $\frac{57}{60}$

Unit 7
Subtraction of Common Fractions

1. $\frac{5}{9}$
3. $4\frac{4}{7}$
5. $6\frac{44}{63}$
7. $\frac{2}{15}$
9. $3\frac{1}{10}$
11. $\frac{3}{8}$ feet
13. $2\frac{1}{4}$ inches
15. $124\frac{3}{4}$ feet
17. $17\frac{3}{4}$ feet
19. $\frac{13}{16}$ kW
21. length: $94\frac{5}{8}$ inches
    width: $45\frac{3}{4}$ inches
23. $4\frac{3}{4}$ feet
25. $1\frac{3}{4}$ hours

Unit 8
Multiplication of Common Fractions

1. $\frac{2}{9}$
3. $1\frac{33}{35}$
5. $\frac{1}{9}$
7. $4\frac{1}{2}$
9. $\frac{1}{6}$ inch
11. $1\frac{1}{2}$ feet
13. 40 feet
15. $2{,}689\frac{1}{6}$ ft
17. $1\frac{9}{16}$ inches
19. $1\frac{9}{20}$ kW
21. $493\frac{1}{2}$ sq ft
23. $6\frac{3}{7}$ kW
25. $96\frac{15}{16}$ pounds

Unit 9
Division of Common Fractions

1. $\frac{20}{33}$
3. $\frac{5}{8}$
5. $\frac{21}{22}$
7. $3\frac{1}{9}$
9. $\frac{1}{16}$ pound
11. $290\frac{17}{45}$ gallons per minute
13. 64 turbines
15. $6\frac{61}{84}$ hours
17. $47\frac{1}{4}$ in
19. $42\frac{6}{7}$ feet
21. $18\frac{2}{11}$ sq ft
23. $86\frac{37}{412}$ watts per gallon
25. $17\frac{1}{2}$ hours

Unit 10
Combined Operations with Common Fractions

1. $1\frac{11}{36}$
3. $\frac{5}{16}$
5. $\frac{2}{5}$
7. $1\frac{1}{2}$
9. $5\frac{9}{35}$ gallons

11. $10\frac{1}{2}$ sq in
13. $3\frac{11}{15}$ kW
15. $178\frac{5}{6}$ ft
17. $5\frac{5}{12}$ ft
19. $31\frac{57}{64}$ sq ft
21. $1,925\frac{1}{10}$ kWh
23. $25\frac{563}{1,114}$ seconds
25. $5\frac{41}{84}$ hours

Section 3
Decimal Fractions

Unit 11
Addition of Decimal Fractions

1. 79.885
3. 151.515
5. 212.500 MW
7. 52.197
9. 446.712
11. 44.3441 m^3
13. 51.35 hours
15. $12,342.10
17. 313.9 ft
19. No. (198.6 ft)
21. 127.08 man-hours
23. 630.25 ft
25. 193.5 ft

Unit 12
Subtraction of Decimal Fractions

1. 43.31
3. 284.077
5. 3,044.59 pounds
7. 32.26
9. 330.529
11. 268.025 hours
13. 175.55 kWh
15. 3.9 man-hours
17. $1,995.65
19. 78.85 meters
21. 503.47 sq ft
23. 446.3 ft
25. 42.5 ft

Unit 13
Multiplication of Decimal Fractions

1. 306.711
3. 0.5030606

5. 1.518542 pounds
7. 676.375 sq ft
9. 2,057.0175
11. $147.81
13. 3,065,428.8 kWh
15. 7,328.8125 pounds
17. $221.05
19. $223.44
21. 148.512 Btu
23. 202.125 feet
25. 3,652,368.48 gallons

Unit 14
Division of Decimal Fractions

1. 26.4
3. 0.0517
5. 741.300 MW
7. 0.00677 ft
9. 0.483
11. 2.04 miles
13. 0.288 kWh
15. $31.20
17. 0.57 ft drop per ft of travel
19. 16 turbines
21. 0.23275 kW
23. 0.03598 gallons
25. 14 poles

Unit 15
Decimal and Common Fraction Equivalents

1. 0.625
3. 0.8
5. 0.15
7. 1/5
9. 3/8
11. 0.667

13. 0.111

15. 0.636

17. 604.8 gpm

19. 416.8125 ft

21. 12.4 lbs

23. 0.19

25. 538.7 ft

Unit 16
Combined Operations with Decimal Fractions

1. 880.36

3. 4.846652

5. 53.434

7. 0.219165

9. 14.310 gpm

11. 7.542233 cu ft

13. 0.333

15. 2.625 ft

17. 527.49 kW

19. 8,657.52 kWh

21. 190 bags

23. 9.84375 hours

25. 2.2275 kW

Section 4
Averages, Ratio, and Proportion

Unit 17
Averages

1. 29
3. 605
5. 199.7
7. 12.7 ft
9. $1,119.09
11. 6,144 kWh
13. 2,542 gpm
15. $1\frac{35}{36}$ hours
17. 28.88 days
19. 78 ft/hr
21. $78.77
23. 38.25 hr
25. $5,771.50

Unit 18
Ratio

1. 5 : 7
3. 7 : 5
5. 11 : 21
7. 0.4 : 1
9. $\frac{21}{62}$: 1

11. 20 : 37
13. 57 : 17
15. 74 : 57
17. 40 : 57
19. 30 : 1
21. 550 ft : 1 ton
23. 1 kW : 317 gpm
25. 376,331 sq ft : 660 ft

Unit 19
Proportion

1. 24
3. $19\frac{1}{5}$
5. $3\frac{2}{3}$
7. 45
9. 5.338
11. X = 6; Y = 25
13. 1,925 ft
15. 1,378.125 W
17. 544 lbs
19. 347.82 ft
21. 2 gpm
23. 3.13 kWh
25. 71.5 hours

Section 5
Percentage, Discount, Markup, and Efficiency

Unit 20
Percentage

1. 0.35
3. 0.025
5. 41%
7. 150%
9. 0.68
11. 262.5
13. 4.2%
15. 37.5%
17. 17.4%
19. $250.00
21. 62.5%
23. 69%
25. 2,400 kW

Unit 21
Discounts and Markups

1. $198.30
3. $223.83
5. $65.46
7. $127.85
9. $598.00
11. $1,647.18
13. $2,494.57

15. $6,379.74
17. $3,803.52
19. $742.50
21. A. $415.17
    B. $408.94
23. A. $3,473.91
    B. $3,527.35
25. A. $1,758.87
    B. $1,884.51

Unit 22
Efficiency

1. 68%
3. 196.8 kW
5. 32%
7. 12.5%
9. 375%
11. 3,326.7 Btu/h
13. 590.1 kW
15. 426.6 kW
17. 230.4 W
19. A. 204 W
    B. 21.6 W
21. 5.4%
23. 226.88%
25. 360%

Section 6
Direct Measure

Unit 23
Equivalent Units of Temperature Measure

1. 149°F
3. 44.6°F
5. 10°C
7. 0°C
9. 25°C
11. 54°F
13. 302°F
15. 42.8°F
17. 185°F
19. $15\frac{5}{9}$°C
21. 36 Fahrenheit degrees
23. $16\frac{2}{3}$ Celsius degrees
25. $27\frac{7}{9}$ Celsius degrees

Unit 24
Angular Measure

1. 48°
3. 140°
5. 90°
7. A = 33°
   B = 15°
   C = 132°
9. A = B = 45°
11. 122°
13. 11°
15. 63°
17. 25°
19. 120°
21. 21°
23. 37°
25. 112°

Unit 25
Units of Length Measure

1. 24 inches
3. 75 inches
5. 300 centimeters
7. 40 centimeters
9. 4 feet
11. $2\frac{11}{12}$
13. 35 meters
15. 0.08 meters
17. 55 inches
19. 658 centimeters
21. $32\frac{1}{2}$ feet
23. $275\frac{5}{6}$ feet
25. 396 inches

Unit 26
Equivalent Units of Length Measure

1. 20.32 centimeters
3. 104.14 centimeters
5. 1.52 meters
7. 6.181 inches
9. 6.988 feet
11. 2 feet 11.827 inch
13. 201.77 feet
15. 754.38 centimeters
17. A. 3,810 centimeters
    B. 1,219.2 centimeters
19. 63.98 feet
21. 7.52 meters
23. 15.75 inches
25. 1.57 inches

Unit 27
Equivalent Units of Additional
Direct Measure

1. 7.57 liters
3. 3.70 gallons
5. 26,400 feet
7. 1.60 miles
9. 19.31 kilometers

11. 151.42 liters per second
13. 10.57 gallons per second
15. 0.38 meters per second
17. 0.04 L/sec
19. 19,440,000 gallons per hour
21. 26.79 kilograms per meter
23. 120.68 kilometers per hour
25. Yes. (141.95 liters per second)

Section 7
Computed Measure

Unit 28
Computed Length Measure

1. 5 ft 4 in
3. 4.71 inches
5. 132 feet 1 inch
7. 3,808 centimeters
9. 596.90 centimeters
11. 24 meters 10 centimeters
13. 151 feet
15. 5 miles 470 feet
17. 353 feet 7 inches
19. 7 feet 5 inches
21. 480.66 feet
23. 0.019 inches
25. 0.256 inches

Unit 29
Area Measure

1. 3,220 square feet
3. 196 square inches
5. 10.18 square meters
7. 150.06 square meters
9. $759\frac{1}{2}$ square feet
11. 588 square feet
13. 4.6225 square meters
15. 67,584 square feet
17. 33 square inches
19. 9,160.91 square meters
21. 155 feet
23. 54.78 square feet
25. 100.53 square yards or 904.77 square feet

Unit 30
Volume Measure

1. 27,048 cubic feet
3. 42.875 cubic feet or $42\frac{7}{8}$ cubic feet
5. 20.80 cubic meters
7. 8.84 cubic feet
9. 2.15 cubic meters
11. 67,500 cubic feet
13. 896 cubic feet
15. 1,344 cubic feet
17. 34.33 cubic meters
19. 10,584 cubic feet
21. 4,433.43 cubic centimeters
23. 166.72 inches
25. 7,040.14 cu in

Unit 31
Equivalent Units of Area and Volume Measure

1. 14.25 acres
3. 4.38 square inches
5. 32,670 square feet
7. 1,095,638.4 square inches
9. 0.30 acres
11. 3,095 panels
13. 39,938.13 watts
15. 6.95 cubic feet/sec
17. 4.58 cubic yards
19. 137.73 cubic feet
21. 0.96 yards
23. 2.20 feet
25. No. (164.03 liters per second)

Section 8
Formulas

Unit 32
Electrical Relationships I

1. 372 volts
3. 15 ohms
5. 311.88 volts
7. 5.02 volts
9. 0.0057 amps
11. 9.2 amps
13. 19.17 ohms
15. 0.81 volts
17. 15.19 ohms
19. 3.79 ohms
21. 27 ohms
23. 209.67 volts
25. A. 6.33 amps
    B. 8.55 volts

Unit 33
Electrical Relationships II

1. 66 volts
3. 1,680 watts
5. 0.30 watts
7. 565.2 watts
9. 0.06 ohms
11. 19,602 watts
13. 0.0012 ohms
15. 230.05 volts
17. 25.23 watts
19. A. 293.88 watts
    B. 2.45 amps
21. 10.87 amps
23. 3.11 amps
25. 1,412.93 amps

Unit 34
Heat Transport

1. 360 Btu/min
3. 56 pounds/min
5. 21°F difference
7. 2,399.25 Btu/min
9. 31.16°F difference
11. 1,047.63 Btu/min
13. 70%
15. 1,430 Btu/min
17. 1.79°F
19. 1,468.5 Btu/min
21. 7,016.1 Btu/hr
23. 643.57 Btu/min
25. 26.15 Btu/min

Unit 35
Energy

1. 60 foot-pounds
3. 14.36 joules
5. 32 foot-pounds
7. 1,536.15 Btu/hr
9. $41\frac{1}{2}$ foot-pounds or 41.5 foot-pounds
11. 368,387.67 watt-hr
13. 150,903.45 Btu
15. 1,585.14 joules
17. 760.87 joules
19. 171,936 watt-seconds (joules)
21. 39.15 Btu/sec
23. 237.85 amps
25. 64,435,072.18 watts

Section 9
Trigonometry

Unit 36
Trigonometric Functions

1. 0.6
3. 1.33
5. 0.45
7. 11.63
9. 8.39
11. 22° is the closest angle in the table. A calculator would give the angle as 21.80°.
13. 14.14
15. 34°
17. 65°
19. 147.75 in or 12 ft 3.75 in
21. 68.82 in or 5 ft 8.82 in
23. 25.04 ft
25. 2.44 ft

Section 10
Graphs

Unit 37
Graphs and Graphing

1.

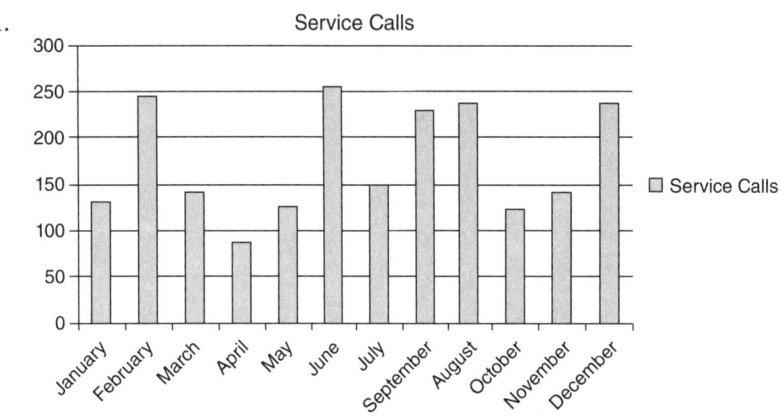

3.   Need 3. None extra.
5.   135°
7.   Max 77°F, min 40°F
9.   About 60 days
11.   About 60 hours
13.   12 mph. 425 hours
15.   50 watts
17.   80 hours
19.   775 watts
21.

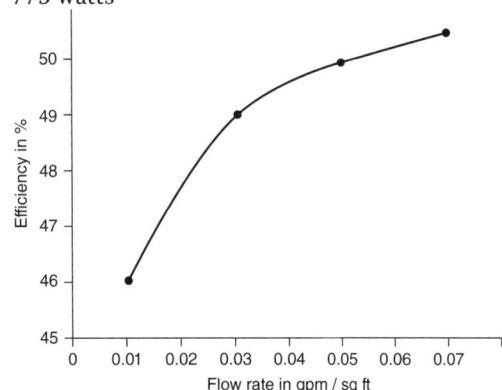

23.   0.9 gpm
25.   60 ft

Section 11
Bills

Unit 38 Estimates and Billing

1.

No.	WE HAVE IT SUPPLY COMPANY		Date:	
	Ourtown, U.S.A.			
Qty	**ITEM**		**Price Per**	**PRICE**
1	4,000 watt PV system		$23,680.00	$23,680.00
1	Electrical equipment and wiring		2,270.00	2,270.00
40	Labour (man-hours)		50.00	2,000.00
	Estimate			
	Good for 1 month			
		TOTAL		$27,950.00

3.

No.	WE HAVE IT SUPPLY COMPANY Ourtown, U.S.A.		Date:	
Qty	**ITEM**		**Price Per**	**PRICE**
3	Gaskets		$18.45	$55.35
75 ft	6" PVC piping		2.25	168.75
5	cans PVC sealant		12.95	64.75
	Delivery charge		45.00	45.00
		TOTAL		$333.85

5.

Qty	ITEM	Price Per	PRICE
	WE HAVE IT SUPPLY COMPANY No. Date: Ourtown, U.S.A.		
1	Pump shaft (New)	$17.45	$17.45
1	can sealant	14.34	14.34
3.5 hr	Labor	35.00	122.50
	Pick up / Drop off Charge	65.00	65.00
	TOTAL		$219.29

7.

No.	WE HAVE IT SUPPLY COMPANY Ourtown, U.S.A.	Date:	
Qty	**ITEM**	**Price Per**	**PRICE**
1	New PV panel	$1,480.00	$1,480.00
	New wiring and connectors	43.75	43.75
4 hr	Labor	47.00	188.00
	Service Call	50.00	50.00
	Subtotal		1,761.75
	7% tax		123.32
	TOTAL		$1,885.07

9.

Qty	ITEM	Price Per	PRICE
	KEEP IT GREEN AND CLEAN COMPANY Yourtown, MD	No.	Date:
2	wells drilled	$7,500.00	$15,000.00
1,300 ft	piping for wells	2.93	3,809.00
135	bags of grout	27.00	3,645.00
26	tons of sand	12.75	331.50
	Parts subtotal		7,785.50
	6% tax		467.13
60 hr	Labor	42.00	2,520.00
	TOTAL		$25,772.63

INDEX